AF736845

Reviews of

102 Physiology, Biochemistry and Pharmacology

Editors
R. H. Adrian, Cambridge · H. zur Hausen, Freiburg
E. Helmreich, Würzburg · H. Holzer, Freiburg
R. Jung, Freiburg · R. J. Linden, Leeds
P. A. Miescher, Genève · J. Piiper, Göttingen
H. Rasmussen, New Haven · U. Trendelenburg, Würzburg
K. Ullrich, Frankfurt/M. · W. Vogt, Göttingen
A. Weber, Philadelphia

With 58 Figures

Springer-Verlag Berlin Heidelberg GmbH

ISBN 978-3-662-31016-8 ISBN 978-3-540-39417-4 (eBook)
DOI 10.1007/978-3-540-39417-4

Library of Congress-Catalog-Card Number 74-3674

This work is subject to copyright. All rights are reserved, whether the whole or part of the material is concerned, specifically those of translation, reprinting, re-use of illustrations, broadcasting, reproduction by photocopying machine or similar means, and storage in data banks. Under § 54 of the German Copyright Law where copies are made for other than private use, a fee is payable to 'Verwertungsgesellschaft Wort', Munich.

© by Springer-Verlag Berlin Heidelberg 1985
Originally published by Springer-Verlag Berlin Heidelberg New York Tokyo in 1985
Softcover reprint of the hardcover 1st edition 1985

The use of registered names, trademarks, etc. in this publication does not imply, even in the absence of a specific statement, that such names are exempt from the relevant protective laws and regulations and therefore free for general use.

Product Liability: The publisher can give no guarantee for information about drug dosage and application thereof contained in this book. In every individual case the respective user must check its accuracy by consulting other pharmaceutical literature.

2127/3130-543210

Contents

Indexed in Current Contents

Rev. Physiol. Biochem. Pharmacol., Vol. 102
© by Springer-Verlag 1985

Relation Between Mitochondrial Calcium Transport and Control of Energy Metabolism

RICHARD G. HANSFORD

Contents

National Institute on Aging, National Institutes of Health, Gerontology Research Center, Baltimore City Hospitals, Baltimore, Maryland 21224, USA

List of Abbreviations

$[Ca^{2+}]c$	Free Ca^{2+} concentration in the cytosol
$[Ca^{2+}]o$	Free Ca^{2+} concentration in the extramitochondrial phase of in vitro experiments
$[Ca^{2+}]m$	Free Ca^{2+} concentration of the mitochondrial matrix
Ca content	The total Ca of a tissue, both bound and free
$\Delta\bar{\mu}/H^+$	The proton electrochemical gradient across the inner mitochondrial membrane
$\Delta\psi$	The membrane potential across the inner mitochondrial membrane
FCCP	Carbonyl cyanide *p*-trifluoromethoxyphenyl hydrazone – an uncoupling agent or proton ionophore
EGTA	Ethyleneglycol-bis-(β-aminoethyl ether)-*N*,*N*¹-tetraacetic acid
Hepes	*N*-2-hydroxyethylpiperazin-*N*¹-2-ethanesulfonic acid
Pi	Orthophosphate, inorganic phosphate

1 Introduction

Mitochondrial Ca^{2+} uptake is a process of great biochemical and physiological interest. It has been shown to be linked directly to the fundamental energy transducing system in the inner mitochondrial membrane, to have the capacity to transport rapidly large amounts of calcium from cytosol to the inner mitochondrial matrix space, and to be ubiquitous in mammalian mitochondria. It has been the subject of a vast amount of research work, much of which is summarized in several recent cogent reviews (*Fiskum* and *Lehninger* 1982; *Nicholls* and *Åkerman* 1982; *Åkerman* and *Nicholls* 1983). In addition to this uptake process, there is a parallel efflux pathway by which Ca^{2+} leaves the mitochondrion, and it has been realized rather recently that there is in fact a continuous cycling of the ion across the mitochondrial membrane. The energetics of this calcium-cycling process and the potential that it provides for control of cellular events have been the subject of a recent review in this series (*Åkerman* and *Nicholls* 1983). The emphasis of their article was on the role that mitochondrial Ca^{2+} transport may play in the regulation of the free Ca^{2+} concentration of the cytosol ($[Ca^{2+}]c$). The present article has an entirely different perspective and puts forward and examines the idea that mitochondrial Ca^{2+} transport serves to regulate the free Ca^{2+} concentration of the mitochondrial matrix ($[Ca^{2+}]m$). This has the potential advantage to the animal that the activity of major catabolic pathways, which are Ca^{2+} sensitive and intramitochondrial, can be adjusted to the energy demands placed upon the tissue and signalled through changes in cytosolic Ca^{2+}. Indeed, a case will be made that in muscle and in nervous tissue the phenomenon of respiratory control, long thought of in terms of the availability of ADP to the mitochondria (*Chance* and *Williams* 1956), should instead be thought of in terms of the availability of both ADP and Ca^{2+} ions. Each of these serves to activate enzymes of the Krebs tricarboxylate cycle, the final common pathway of catabolism, and thus to help the working cell maintain the adenine nucleotide phosphate potential and allow efficient performance of cellular work. Further, the possibility exists that control of catabolism by Ca^{2+} may function independently of control by ADP, allowing mitochondria in hormone-sensitive tissue to accelerate energy transduction in response to a hormone which elevates $[Ca^{2+}]c$, without the need to disturb the cellular phosphate potential. This possibility will be developed in some detail below. Similar views on mitochondrial Ca^{2+} transport have been expressed in excellent short reviews by *Denton* and *McCormack* (1980, 1981).

The model to be presented requires that each of the following conditions be satisfied:

1. Intramitochondrial enzymes catalyzing rate-limiting reaction steps in catabolic pathways are activated by Ca^{2+}.
2. Values of $[Ca^{2+}]c$ increase in response to conditions requiring higher rates of energy transduction in the mitochondrion.
3. Transport processes exist in the mitochondrial membrane which are capable of raising $[Ca^{2+}]m$ in response to the change in the cytosol. This requires a net inward flux of Ca^{2+} across the mitochondrial membrane, which must reverse when the change in the $[Ca^{2+}]c$ reverses.
4. The values of $[Ca^{2+}]m$ occurring in vivo must be in the range where such changes elicit changes in enzyme activity, i.e., the model would be invalidated if these concentrations were so high as to continually saturate the Ca^{2+}-sensitive intramitochondrial enzymes.

Each of these assumptions will be scrutinized in turn. In addition, the separate possibility will be considered that $[Ca^{2+}]m$ can be altered without the necessity of a prior change in $[Ca^{2+}]c$, if the relative balance of the activities of the mitochondrial uptake and release pathways is altered as might occur in the response of a cell to a hormone.

Finally, this article will examine the idea, recently reviewed and supported by *Fiskum* and *Lehninger* (1982) and *Åkerman* and *Nicholls* (1983), that mitochondria accurately buffer $[Ca^{2+}]c$. Although not necessarily the antithesis of the thesis developed in the current article, such buffering may not be compatible with the concept that changes in $[Ca^{2+}]m$ regulate the activity of intramitochondrial enzymes. This arises because values of $[Ca^{2+}]m$ allowing enzyme regulation are not sufficiently high to allow accurate buffering of $[Ca^{2+}]c$ by the mitochondria, i.e., mitochondria can buffer $[Ca^{2+}]c$ or allow the control by Ca^{2+} of intramitochondrial enzymes, but they cannot do both. The question of which paradigm better describes the behavior of mitochondria in the animal can only be answered by a knowledge of the Ca content of the mitochondria in vivo. The current state of our knowledge of this important parameter will also be discussed.

2 The Ca^{2+}-Sensitive Dehydrogenases

This section discusses evidence that Ca^{2+} ions activate key catabolic enzymes. The first to be considered will be three intramitochondrial dehydrogenases which catalyze nonequilibrium, and thus rate-limiting, reactions in the terminal oxidations of carbohydrates and fatty acids. Following this, other Ca^{2+}-sensitive enzymes which are intramitochondrial but which appear to respond to higher concentrations of Ca^{2+} will be discussed. Finally, the enzyme glycerol 3-phosphate dehydrogenase, which is Ca^{2+} sensitive and forms an integral part of the inner mitochondrial membrane, will be considered.

2.1 Pyruvate Dehydrogenase Phosphate Phosphatase

The pyruvate dehydrogenase complex catalyzes a nonequilibrium reaction committing carbon from carbohydrate to the formation of fat or to complete oxidation to CO_2. Part of the metabolic control of this enzyme is achieved by a phosphorylation/dephosphorylation cycle, catalyzed by a kinase and a phosphatase respectively. The dephosphoenzyme (PDH_A) is the catalytically active form. These relations have been reviewed recently by *Hansford* (1980), *Reed* (1981), and *Wieland* (1983). *Denton* et al. (1972) discovered that the phosphatase activity was enhanced by micromolar concentrations of Ca^{2+} and this has been attributed to a decrease in

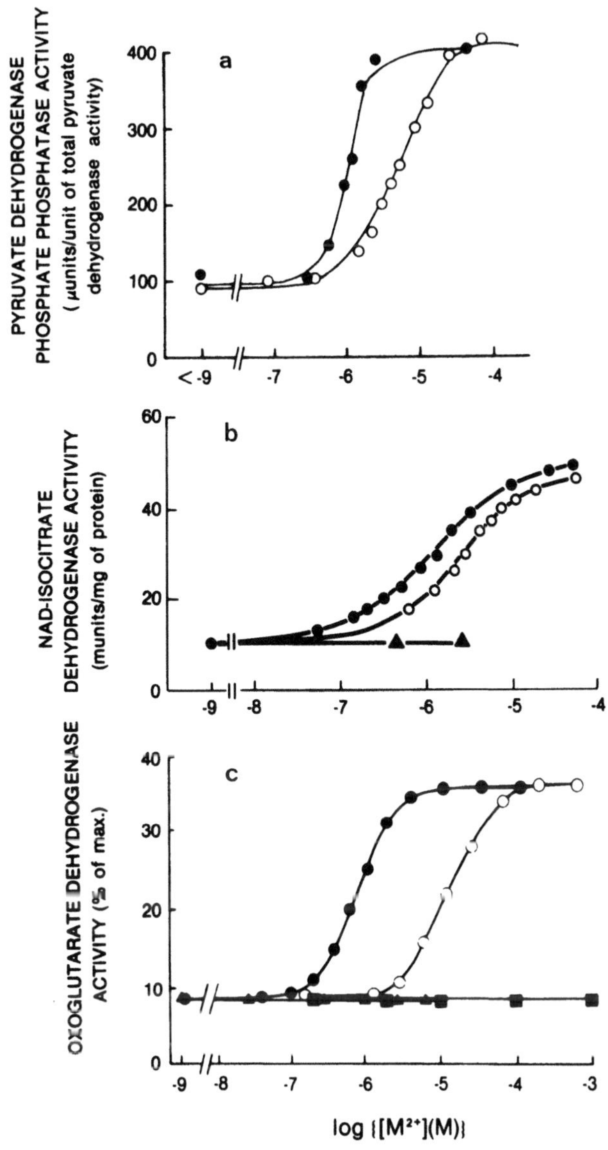

Fig. 1a–c. Activation of **a** pyruvate dehydrogenase phosphate phosphatase, **b** NAD-isocitrate dehydrogenase, and **c** 2-oxoglutarate dehydrogenase by Ca^{2+} (●) and Sr^{2+} (○) ions. Enzyme activity was measured in mitochondrial extracts (**a,b**) or with the purified protein (**c**), as described in the original papers. The indicated values of $[Ca^{2+}]$ and $[Sr^{2+}]$ were maintained with EGTA buffers. (**a**, *McCormack* and *Denton* 1980; **b**, *Denton* et al. 1978; **c**, *McCormack* and *Denton* 1979)

the *K*m of the enzyme for its substrate, pyruvate dehydrogenase phosphate (*Pettit* et al. 1972; *Randle* et al. 1974). Although Ca^{2+} causes a change in the affinity of the enzyme for its substrate, as in the other cases to be discussed, the net effect is the generation of more PDH_A and thus an increased V_{max} of the PDH complex.

It is seen in Fig. 1 that the $K_{0.5}$ for Ca^{2+} activation of the isolated phosphatase is approximately 1 μM. These data are derived from studies employing the phosphatase of white adipose tissue mitochondria, but the sensitivity of the enzyme from heart is similar (*Denton* et al. 1975). However, there is also the possibility that the pyruvate dehydrogenase kinase from heart is inhibited by Ca^{2+} with approximately the same sensitivity (*Cooper* et al. 1974): this would be expected to result in a still steeper dependence of PDH_A content on $[Ca^{2+}]m$ in the heart. Presented together with the sensitivity of the phosphatase to Ca^{2+} in Fig. 1 is the sensitivity to Sr^{2+}, which has been found by *Denton* and coworkers to activate the trio of Ca^{2+}-sensitive intramitochondrial dehydrogenases. Activation by Sr^{2+} requires a concentration approximately one order of magnitude higher than that needed for activation by Ca^{2+} (Fig. 1).

2.2 NAD-Isocitrate Dehydrogenase

The enzyme NAD-isocitrate dehydrogenase catalyzes a nonequilibrium reaction in the tricarboxylate cycle and is activated by a decline in the ratios ATP/ADP and $NADH/NAD^+$ (see *Hansford* 1980, for a review). The discovery that NAD-isocitrate dehydrogenase is also activated by micromolar concentrations of Ca^{2+} (*Denton* et al. 1978) raised the intriguing possibility that flux through this reaction could be adjusted to the demands of the work load not only by the availability of ADP to the mitochondria but also by the availability of Ca^{2+}.

Again, the $K_{0.5}$ for stimulation by Ca^{2+} is approximately 1 μM in experiments with mitochondrial extracts (Fig. 1), with Sr^{2+} being effective at somewhat higher concentrations. Ca^{2+} acts as an allosteric activator, decreasing the apparent K_m of the enzyme for its substrate isocitrate (Fig. 2). These observations have been extended recently to the purified enzyme from bovine heart, with the finding that the $K_{0.5}$ value for Ca^{2+} decreases with increasing concentrations of the substrate, magnesium isocitrate, with the limiting value of the $K_{0.5}$ for Ca^{2+} being 0.29 μM (*Aogaichi* et al. 1980). The apparent *K*m for isocitrate is decreased sharply with decreasing pH and with increasing ADP concentration, such that a wide range of values is obtained under different experimental conditions. These changes have been tied to changes in aggregation of the enzyme subunits, at least for the effect of adenine nucleotides (*Giorgio* et al. 1970). It is noteworthy that Ca^{2+} was found to be an activator only in the presence of ADP (*Denton* et al. 1978; *Aogaichi* et al. 1980), lending credence to the idea of respiratory control by ADP plus Ca^{2+}. It would be interesting

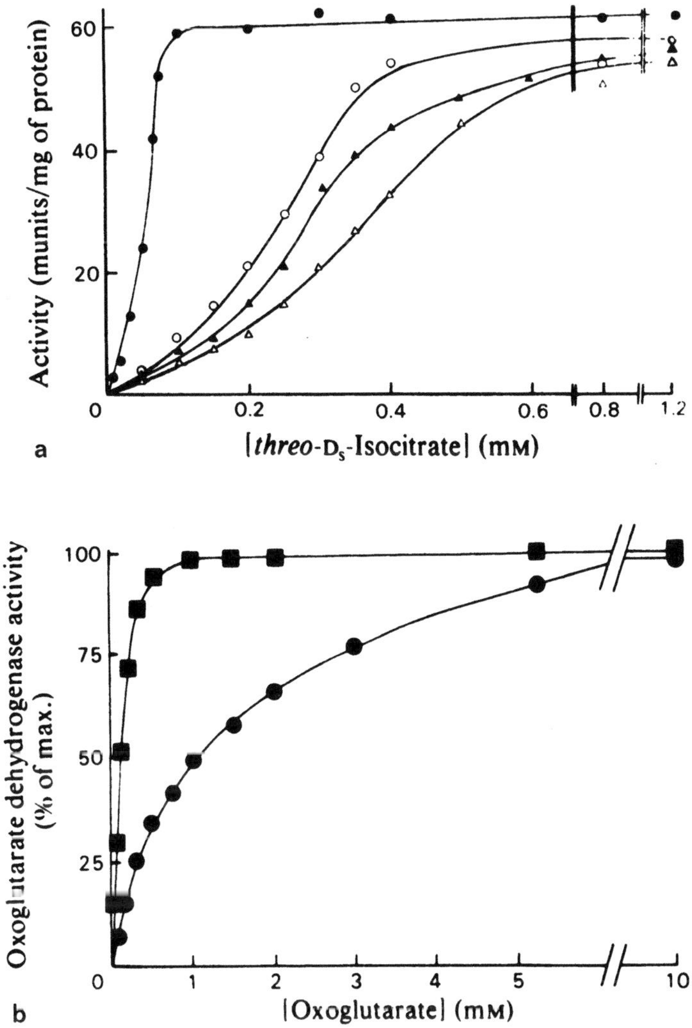

Fig. 2a,b. Effect of Ca^{2+} on the apparent substrate affinity of **a** NAD-isocitrate dehydrogenase and **b** 2-oxoglutarate dehydrogenase. Enzyme activity was determined using a mitochondrial extract (**a**) or the purified protein (**b**) as described in the original papers. Symbols: **a** ▲, 5 m*M* EGTA; △, 5 m*M* EGTA plus 5 m*M* $CaCl_2$ ([Ca^{2+}] approx. 30 μM); ○, 5 m*M* EGTA plus 1 m*M* ADP; ●, 5 m*M* EGTA plus 5 m*M* $CaCl_2$ plus 1 m*M* ADP ([Ca^{2+}] approx. 30 μM); **b** ●, 5 m*M* EGTA; ■, 5 m*M* EGTA + 5 m*M* $CaCl_2$ ([Ca^{2+}] approx. 30 μM). (**a**, *Denton* et al. 1978; **b**, *McCormack* and *Denton* 1979)

to know whether the enzyme shows Ca^{2+} sensitivity at ATP/ADP ratios plausible for the mitochondrial matrix (around 1–4, see *Davis* and *Lumeng* 1975; *Akerboom* et al. 1978), in view of the proposal that hormonal stimulation may be mediated via changes in [Ca^{2+}]c in the absence of changes in ATP/ADP ratio (*Denton* and *McCormack* 1980, 1981; and see below).

2.3 2-Oxoglutarate Dehydrogenase

The enzyme 2-oxoglutarate dehydrogenase catalyzes the remaining clearly nonequilibrium reaction of the tricarboxylate cycle and is the primary site of control for the segment of the tricarboxylate cycle from 2-oxoglutarate to malate (*Randle* et al. 1970; *Smith* et al. 1974; see *Williamson* 1979 for review). In remarkable analogy to their findings with NAD-isocitrate dehydrogenase, *McCormack* and *Denton* (1979) found that the 2-oxoglutarate dehydrogenase complex from pig heart, rat heart, and rat brown adipose tissue is activated by Ca^{2+}, with a $K_{0.5}$ of 1.2 μM (Fig. 1). As with NAD-isocitrate dehydrogenase, Sr^{2+} activates, but at higher concentrations. The effect of Ca^{2+} is to decrease the apparent K_m for the substrate 2-oxoglutarate without affecting that for NAD^+. The enzyme is also activated by a decrease in pH and a decrease in the ATP/ADP ratio (*Hansford* 1972; *McCormack* and *Denton* 1979). However, although Ca^{2+}, H^+, and adenine nucleotides each act to alter the apparent K_m for 2-oxoglutarate, the effects seem independent of each other, at least over the concentration range studied (*McCormack* and *Denton* 1979). Thus, the value for the apparent K_m for 2-oxoglutarate is several hundredfold higher in the presence of ATP and absence of Ca^{2+} than it is in the presence of ADP and presence of Ca^{2+}, at pH 7.2 (*McCormack* and *Denton* 1979).

The extreme sensitivity of 2-oxoglutarate dehydrogenase to Ca^{2+} became clearer with the finding of *Lawlis* and *Roche* (1980) that Ca^{2+} relieves inhibition by NADH, as well as decreasing the apparent K_m for 2-oxoglutarate as discussed above. The former effect may be quantitatively more important. Thus, *Lawlis* and *Roche* (1980) reported an 18-fold increase in the activity of the 2-oxoglutarate dehydrogenase complex over the range of Ca^{2+} concentration $10^{-9}-10^{-5}$ M in the presence of NADH, compared to only a 2.2-fold increase in its absence. The interaction between effects of Ca^{2+} and of NADH is shown in Fig. 3, from the work of *Lawlis* and *Roche* (1981b) on the 2-oxoglutarate dehydrogenase from kidney. It is seen that there is a pronounced sensitivity to Ca^{2+} when the $NADH/NAD^+$ ratio is in the physiological range (log $NADH/NAD^+$ approximately –0.5–0.5; see *Hansford* and *Castro* 1981). Moreover, this experiment was so designed as to remove effects of Ca^{2+} on the apparent K_m for 2-oxoglutarate, in that a higher 2-oxoglutarate concentration was used in the experiments lacking Ca^{2+}, such that activities were exactly matched for all four experimental conditions in the absence of added NADH. During the physiological transition from rest to work in a muscle the 2-oxoglutarate concentration may rise or change little (*Neely* et al. 1972; *Johnson* and *Hansford* 1975), $[Ca^{2+}]c$ rises leading to an increase in $[Ca^{2+}]m$ (see below), and the cytosolic ATP/ADP ratio falls slightly, giving rise to a decrease in the mitochondrial $NADH/NAD^+$ ratio (*Illingworth* et al.

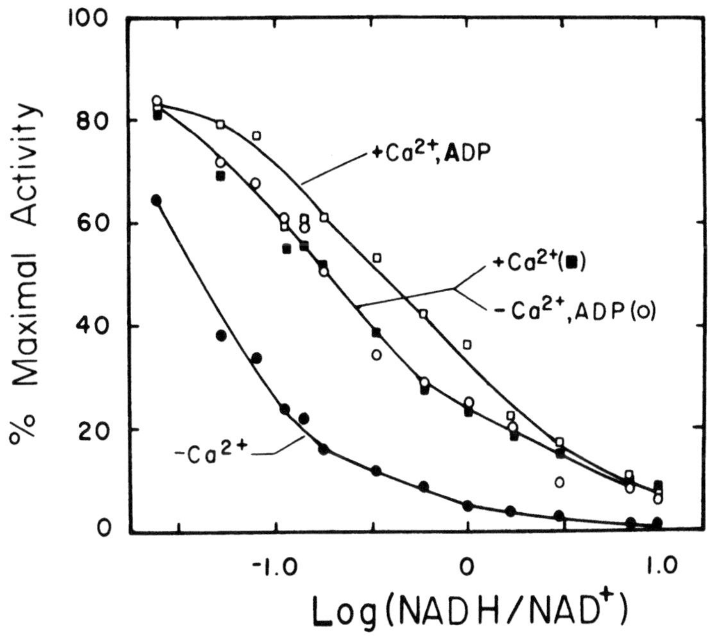

Fig. 3. Effect of Ca^{2+} and ADP on the activity of 2-oxoglutarate dehydrogenase at a variety of $NADH/NAD^+$ ratios. The enzyme used was the purified complex from bovine kidney. The sum of NAD^+ and NADH was 0.5 m*M* and the 2-oxoglutarate concentration was adjusted to give equal control (100%) activities, i.e., minus NADH, under the different conditions of Ca^{2+} and ADP availability. Thus, the effects of Ca^{2+} and ADP seen here reflect a lessening of inhibition by NADH and not a change in $S_{0.5}$ for 2-oxoglutarate. For further details, see the original paper. (*Lawlis* and *Roche* 1981b)

1975; *Hiltunen* and *Hassinen* 1976; *Nishiki* et al. 1978). As both an increased $[Ca^{2+}]m$ and a decreased mitochondrial ATP/ADP ratio will decrease the apparent K_m for 2-oxoglutarate (*McCormack* and *Denton* 1979; *Lawlis* and *Roche* 1981a), and reverse the inhibition by NADH (*Lawlis* and *Roche* 1981b), the net increase in flux through this reaction could be very large in working as compared with resting muscle. The relationship of changes in ATP/ADP ratio and $[Ca^{2+}]c$ in the kidney and other nonexcitable tissues is not so clear as it is in muscle. In these latter tissues, it is possible that activation of the enzyme complex by one of these modulators may occur independently of the others.

Mechanistically, NADH inhibits through binding to a site on the El (or decarboxylase) subunit of the 2-oxoglutarate dehydrogenase complex, a conclusion reached by *Lawlis* and *Roche* (1981b) using a resolved complex and glyoxylate as an artificial acceptor. This is also consistent with the nonequilibrium nature of this partial reaction. The degree of NADH inhibition is diminished by the presence of the Ca^{2+} ion.

2.4 Other Ca^{2+}-Sensitive Intramitochondrial Enzymes

The branched-chain α-ketoacid dehydrogenase complex (EC 1.2.4.4) is analogous in composition and function to the pyruvate dehydrogenase complex and similarly appears to be regulated by a phosphorylation process (*Patel* and *Olson* 1982; *Paxton* and *Harris* 1982). Recently, *Patel* and *Olson* (1982) provided evidence that Ca^{2+} and Mg^{2+}, present together, result in an elevated steady-state content of active branched-chain α-ketoacid dehydrogenase in respiring rat liver mitochondria. Moreover, Ca^{2+} and Mg^{2+} greatly increased the rate of reactivation of the complex, following a period during which the phosphatase was inactivated. This result was interpreted in terms of an activation of a phosphatase by Ca^{2+}. Though provocative, these results do not strictly allow one to decide whether a kinase or a phosphatase is the site of action of Ca^{2+}, as both enzymes are active to some degree in both experiments. Moreover, 2 m*M* $CaCl_2$ was used, in the presence of the ionophore A23187, so that it is not possible to know whether this enzyme system is responding to micromolar concentrations of Ca^{2+} similar to those which are effective in activating the enzymes discussed above.

There is also recent evidence which shows that Ca^{2+} inhibits the phosphorylation of purified branched-chain α-ketoacid dehydrogenase from liver, with a $K_{0.5}$ for Ca^{2+} of 0.08–0.18 μ*M* (*Paxton* and *Harris* 1982). The effect is small in magnitude, being only 15%. There is thus the possibility that the Ca^{2+} ion acts both to inhibit the kinase and to activate the phosphatase, as found for the pyruvate dehydrogenase system in heart (*Cooper* et al. 1974). These combined effects would be expected to give rise to a sharp dependence of active dehydrogenase upon $[Ca^{2+}]m$. To define the system more clearly, however, it will be necessary to examine the concentration dependence of the activation of the phosphatase by Ca^{2+}, which was inferred to occur in the study by *Patel* and *Olson* (1982).

An activation of several other intramitochondrial dehydrogenases or dehydrogenase systems by Ca^{2+} has also been described but in none of these cases is the effect as well defined as in the examples above. In addition, there is some doubt about whether the concentrations of Ca^{2+} required are physiologically relevant. Thus, *Ezawa* and *Ogata* (1977, 1979) have described a Ca^{2+}-induced activation of succinate dehydrogenase and succinate oxidase in intact rat liver mitochondria. The effect is sharply distinguished from the Ca^{2+} activation of pyruvate dehydrogenase phosphatase, NAD-isocitrate dehydrogenase, and 2-oxoglutarate dehydrogenase occurring in intact respiring mitochondria (this is described below). Activation of succinate oxidation requires higher extramitochondrial Ca^{2+} concentrations than is the case for these enzymes, with a $K_{0.5}$ of 1 μ*M* Ca^{2+} in coupled respiring mitochondria. In addition, there is a time lag of several minutes before the onset of activation, and activation is accompanied by a simultaneous and severe inhibition in the oxidation of NAD-linked substrates (*Ezawa* and *Ogata* 1979). The authors suggest a conformational change of the membrane which selectively activates succinate oxidation. This conclusion is based on observed changes in light scattering. It seems that activation of phospholipases is another possible mechanism, and it would be interesting to see whether the activation of succinate dehydrogenase can be successfully reversed in these experiments, this being a sine qua non of a metabolic control process.

An activation of the β-oxidation of fatty acids by Ca^{2+} has also been proposed on the basis of experiments with hepatocytes (*Otto* and *Ontko* 1974) and isolated liver mitochondria (*Otto* and *Ontko* 1978). Activated oxidation was evident from enhanced rates of ketone body production and O_2 uptake and an increase in the mitochondrial $NADH/NAD^+$ ratio. The site of action of Ca^{2+} could be inferred to be one of the enzymic steps of the β-oxidation spiral, on the grounds that activation was seen with both mid-chain length and long-chain fatty acids. It is hard to relate this phenomenon to the activation of pyruvate dehydrogenase phosphatase, NAD-isocitrate dehydrogenase, and 2-oxoglutarate dehydrogenase, in that much larger loads of Ca were necessary to activate fatty acid oxidation, 80 nmol Ca/mg mitochondrial protein being approximately optimal. However, it would be premature to dismiss these findings as sequellae of the energy drain of massive Ca^{2+} accumulation, as both ATP content and $NADH/NAD^+$ ratio were found to be elevated. This lends credence to the idea of dehydrogenase-level activation. As yet, however, the precise enzymic locus of the effect remains obscure, and indeed the reviewer has had difficulty in reproducing the result in his laboratory (unpublished observations).

Malmström and *Carafoli* (1976) have reported a requirement for Ca^{2+} for the activation of 3-hydroxybutyrate by heart mitochondria. They have suggested that Ca^{2+} potentiates the interaction of the enzyme with an intramembrane pool of NAD, on the grounds that the effect is not seen with inverted submitochondrial particles prepared by sonication. It is not clear that variation of mitochondrial Ca content over the range described below as controlling PDH_A content and flux through the 2-oxoglutarate dehydrogenase reaction in heart mitochondria would have any effect on 3-hydroxybutyrate oxidation, which appears to require Ca contents in excess of 10 nmol/mg mitochondrial protein (*Malmström* and *Carafoli* 1976).

Finally, an increase in the Ca content of liver mitochondria is found to inhibit flux through pyruvate carboxylase (*Foldes* and *Barritt* 1977) and carbamoylphosphate synthetase (*Meijer* et al. 1981) in isolated mitochondria. Thus, there is the possibility that the anabolic pathways of which these enzymes form a part are inhibited by increased mitochondrial Ca in vivo, whereas catabolic pathways are activated, as described above. Although these results are provocative, the range of mitochondrial Ca contents shown to effect changes in enzyme activity is in the case of both pyruvate carboxylase and carbamoylphosphate synthetase somewhat higher (10 nmol/mg mitochondrial protein) than the ranges to which the tricarboxylate cycle enzymes are sensitive, although these values have only been established in heart (see Sect. 3.3). Probably judgement should be suspended until it is shown that fluxes through pyruvate carboxylase and carbamoylphosphate synthetase are inhibited coincidentally with the activation of pyruvate dehydrogenase interconversion and the tricarboxylate cycle, when intact liver mitochondria are incubated in media such that the extramitochondrial Ca^{2+} concentration ($[Ca^{2+}]o$) is in the range of 10^{-6} – 10^{-7} *M* (see below).

2.5 Glycerol 3-Phosphate Dehydrogenase

All of the enzymes discussed above are intramitochondrial and $[Ca^{2+}]m$ sensitive, and thus are regulated by $[Ca^{2+}]c$ only indirectly. By contrast, glycerol 3-phosphate dehydrogenase, which was the first dehydrogenase to be shown to be activated by submicromolar concentrations of Ca^{2+} (*Hansford* and *Chappell* 1967), appears to face outwards on the inner mitochondrial membrane, as judged from the accessibility of the flavin prosthetic group to ferricyanide and the apparent impermeability of the mitochondrial inner membrane to the substrate, glycerol 3-phosphate (*Klingenberg* and *Buchholz* 1970). Thus, this enzyme probably senses $[Ca^{2+}]c$ directly. The mitochondrial glycerol 3-phosphate dehydrogenase catalyzes the oxidation of glycerol 3-phosphate to dihydroxyacetonephosphate. This reaction forms part of a redox shuttle whereby glycolytically-derived reducing power is oxidized in some tissues, notably insect flight muscle (see *Sacktor* 1975, for a review). The activation of this shuttle in fly flight muscle (*Hansford* and *Chappell* 1967) by values of $[Ca^{2+}]o$ (Fig. 4) identical to those required for the activation of actomyosin ATPase (*Chaplain* 1967) makes it quite likely that Ca^{2+} acts as a signal to balance rates of ATP utilization and provision.

Calcium sensitivity of this enzyme has also been shown in mitochondria from mammalian lung (*Fisher* et al. 1973) and liver (*Wernette* et al. 1981). In these cases, a role in a redox shuttle is less clear in view of the low activity of cytosolic NAD–glycerol 3-phosphate dehydrogenase.

The important features of the activation of FAD–glycerol 3-phosphate dehydrogenase by Ca^{2+} are that the $K_{0.5}$ for Ca^{2+} is around 5×10^{-7} *M* (Fig. 4), and that the metal ion acts as an allosteric effector, decreasing the

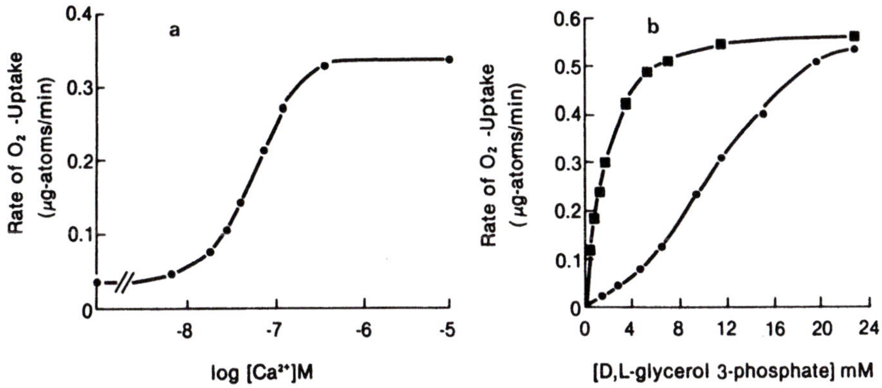

Fig. 4a,b. The activation of mitochondrial glycerol 3-phosphate dehydrogenase by Ca^{2+}. Enzyme activity was followed indirectly by measuring the rate of O_2 uptake of suspensions of blowfly flight muscle mitochondria with glycerol 3-phosphate as substrate, in the presence of ADP or FCCP. **a** Enzyme activity as a function of $[Ca^{2+}]o$, as maintained with EGTA buffers. **b** Enzyme activity as a function of substrate concentration in the presence of EGTA (●) ($[Ca^{2+}] < 10^{-8}$ *M*) and in the presence of 10^{-6} *M* Ca^{2+} (■). (Redrawn from *Hansford* and *Chappell* 1967)

K_m for the substrate without changing the V_{max} (Fig. 4). Interestingly, there is no evidence for an involvement of calmodulin in the Ca^{2+} activation of FAD–glycerol 3-phosphate dehydrogenase, as shown by the persistence of Ca^{2+} sensitivity after electrophoresis of the Triton-solubilized enzyme in the presence of EDTA (*Wernette* et al. 1981).

Thus, mitochondria posses both internal and externally facing dehydrogenases which are activated by Ca^{2+} over the concentration range of $10^{-6}-10^{-7}$ *M*. Subsequent sections of this article will examine the evidence that values of both $[Ca^{2+}]m$ and $[Ca^{2+}]c$ fall within this range, and that changes in $[Ca^{2+}]m$ and $[Ca^{2+}]c$ under physiological circumstances probably play important roles in regulating cellular energy metab olism and in coupling such changes to changes in the work functions of the cell.

3 Ca^{2+} Sensitivity of Dehydrogenase Activity in Intact Mitochondria

3.1 Dependence on $[Ca^{2+}]o$

In the previous section we have discussed the strong evidence that micromolar levels of Ca^{2+} activate at least three intramitochondrial and potentially rate-limiting dehydrogenases involved in energy metabolism. The question arises as to whether this control operates in intact mitochondria, with the enzymes exposed to an ionic and protein environment difficult to reconstruct in the cuvette. This question was first approached using

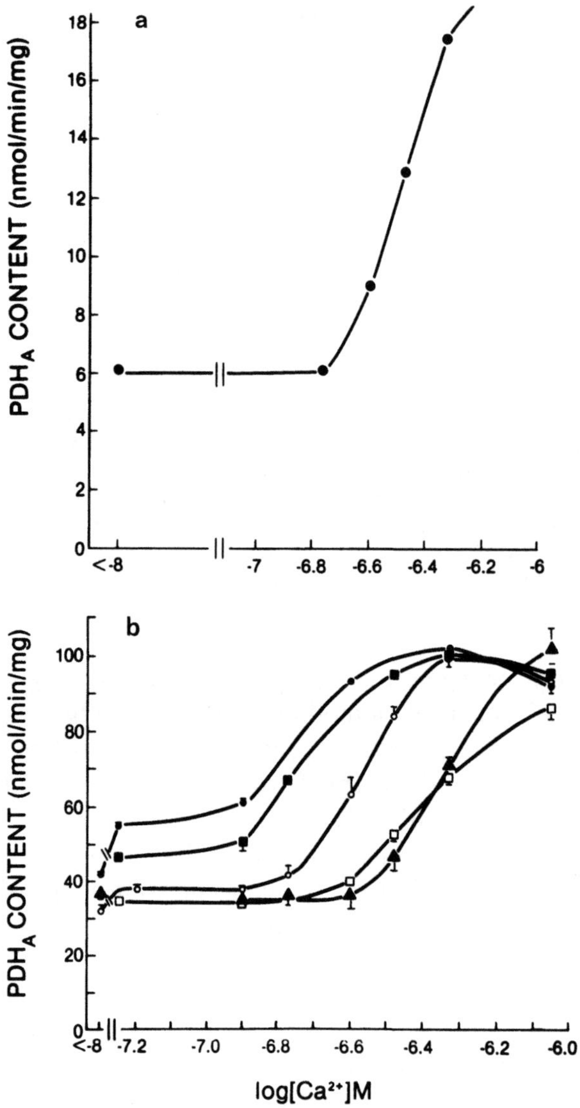

Fig. 5a,b. Effect of variation in $[Ca^{2+}]$o on the PDH_A content of respiring coupled mitochondria from **a** brain and **b** heart. Mitochondria were incubated in media containing the indicated $[Ca^{2+}]$, stabilized with EGTA buffers, and sampled for PDH_A content when enzyme interconversion had reached a steady state. **a** Mitochondria were from rat brain and were incubated in a medium comprising 0.13 *M* KCl, 20 m*M* K Hepes, 5 m*M* KPi, 10 m*M* NaCl, 1 m*M* $MgCl_2$, 5 m*M* succinate, 1 m*M* ATP, 2 mg/ml bovine serum albumin and a Ca^{2+}: EGTA buffer at a concentration of 2 m*M* (as EGTA) and giving the indicated $[Ca^{2+}]$. (Unpublished work of *Austin, Castro,* and *Hansford*). **b** Mitochondria were from rat heart and were oxidizing glutamate plus malate; full details are given in the original paper. Concentrations of Na^+ in the incubations were: ●, none; ■, 0.1 m*M*; ○, 1 m*M*; □, 10 m*M*; ▲, 20 m*M*. All incubations contained 1 m*M* Mg^{2+}. (Redrawn from *Hansford* 1981)

mitochondria treated with the ionophore A23187. The data obtained showed that supplementation of ionophore-treated mitochondria with millimolar concentrations of Mg^{2+} and Ca^{2+} would sustain pyruvate dehydrogenase activity (*Severson* et al. 1974; *Walajtys* et al. 1974). In these studies, the protein-protein interactions of the matrix were maintained, but the normal pathways of Ca^{2+} transport were circumvented. Hence, the results gave no guide to the Ca^{2+} sensitivity of these enzymes in their intramitochondrial locale. However, simple studies in which heart mitochondria were incubated with substrate and without ionophores or uncoupling agents showed that PDH_A content could be greatly diminished by the addition of the chelating agent EGTA (*Hansford* and *Cohen* 1978). Moreover, the maintenance of $[Ca^{2+}]$o with Ca^{2+}-EGTA buffers showed a $K_{0.5}$ for $[Ca^{2+}]$o of between 0.1 and 1 μM for pyruvate dehydrogenase interconversion in these intact mitochondria. Further, removal of extramitochondrial Ca^{2+} with an excess of EGTA resulted in an inactivation of pyruvate dehydrogenase; thus, the effect of Ca^{2+} was shown to be reversible.

These experiments have been refined and extended to the 2-oxoglutarate dehydrogenase in subsequent work by *Denton's* group and by *Hansford*, and some of these results will be illustrated here. Thus, Fig. 5 shows that the PDH_A content of respiring coupled brain and heart mitochondria increases with increases in $[Ca^{2+}]$o over the range 0.1 to 1 μM. Further, the data obtained with heart mitochondria (Fig. 5b) show that the presence of Na^+ has the effect of raising the $K_{0.5}$ for $[Ca^{2+}]$o. Equally, in entirely analogous experiments in which the activity of 2-oxoglutarate dehydrogenase within heart mitochondria is monitored indirectly by studying the fluorescence of mitochondrial NADH (Fig. 6), the inclusion of Mg^{2+} also raises the $K_{0.5}$ for extramitochondrial Ca^{2+}. These results are readily intelligible if enzyme activation is sensitive to $[Ca^{2+}]$m, and if this is maintained by cycling of Ca^{2+} across the inner mitochondrial membrane and is thus coupled to $[Ca^{2+}]$o, as described by *Stucki* and *Ineichen* (1974) and *Nicholls* (1978) and schematized in Fig. 7. Magnesium ions, by inhibiting the electrogenic uniport (*Crompton* et al. 1976a) will diminish the concentration gradient of Ca^{2+} ($[Ca^{2+}]$m > $[Ca^{2+}]$c) across the mitochondrial membrane in the steady state: Na^+ ions, by promoting the Ca^{2+}/n Na^+ antiport system (*Crompton* et al. 1976b; *Crompton* and *Heid* 1978), will have the same effect. When Na^+ concentration is 10–15 mM and Mg^{2+} concentration is 0.5–1 mM, these being physiologically appropriate ranges for heart cytosol (*Veloso* et al. 1973; *Lee* and *Fozzard* 1975), the $K_{0.5}$ for the activation of the enzyme by $[Ca^{2+}]$o was found to be 0.46 μM by *Denton* et al. (1980) and 0.41 μM by *Hansford* (1981). The agreement between these values is quite encouraging.

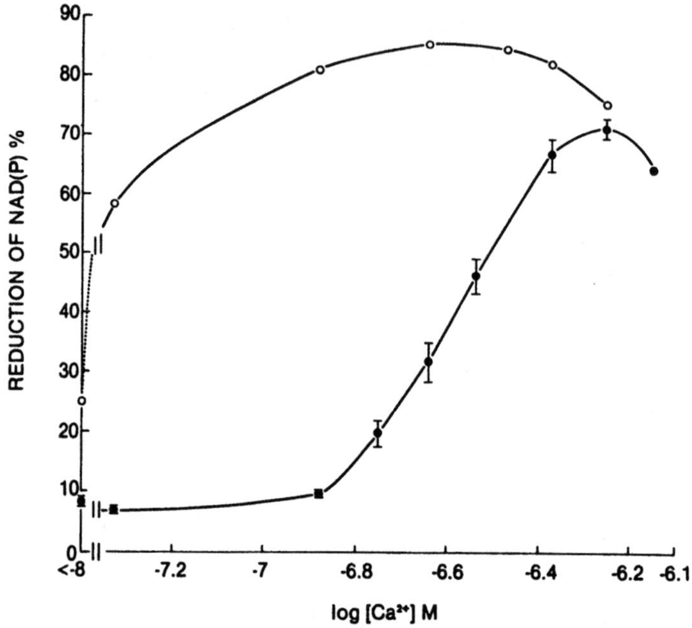

Fig. 6. Dependence of 2-oxoglutarate dehydrogenase activity in respiring, coupled heart mitochondria upon the [Ca^{2+}] of the incubation medium. 2-Oxoglutarate dehydrogenase activity was followed indirectly, by measuring the degree of reduction of mitochondrial NAD by nonsaturating concentrations of 2-oxoglutarate in state 4. Ca^{2+}:EGTA buffers were used to stabilize the values of [Ca^{2+}]o which are indicated. Full details are given in the original paper. The two curves differ in that Mg^{2+} was present at 1 m*M* (●), or absent (○). (Redrawn from *Hansford* and *Castro* 1981)

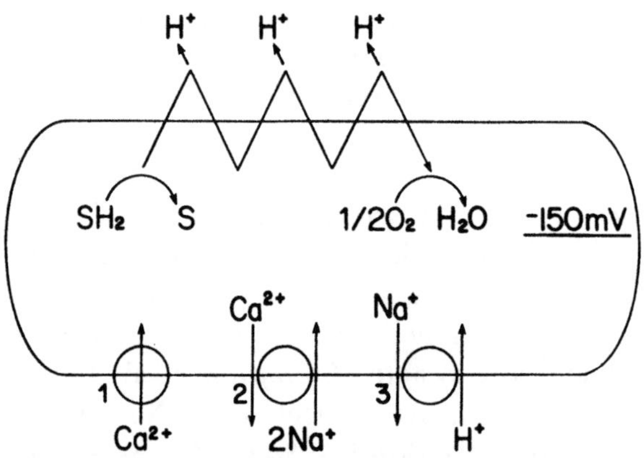

Fig. 7. Scheme of mitochondrial Ca^{2+} transport. Ca^{2+} uptake is electrophoretic, driven by the $\Delta\psi$ generated by respiration, and catalyzed by a uniporter (*1*). Ca^{2+} release requires the participation of an antiporter (*2*), which catalyzes the electroneutral exchange of Na^+ and Ca^{2+} in heart and some other tissues. A Na^+/H^+ antiporter (*3*) completes the circuits for Na^+ and for H^+. For more detail on these relations, see *Fiskum* and *Lehninger* (1982), *Åkerman* and *Nicholls* (1983) and the text

Especially significant is the fact that *Denton* et al. (1980) found a very close correspondence between the $K_{0.5}$ for $[Ca^{2+}]o$ for pyruvate dehydrogenase interconversion and the value for the activation of state-3 oxidation of 2-oxoglutarate. Moreover, comparison of these $K_{0.5}$ values for extramitochondrial Ca^{2+} with the $K_{0.5}$ values obtained with the solubilized enzyme systems allowed the statement that $[Ca^{2+}]m$ is only 2–3 times greater than $[Ca^{2+}]o$ under these conditions (*Denton* et al. 1980). The correspondence between the values obtained for the two different enzymes lends some support to the idea that they both sense a homogeneous matrix Ca^{2+} pool. As is evident from Figs. 5 and 6, a value of 2–3 for the Ca^{2+} gradient across the mitochondrial membrane is appropriate only at extramitochondrial Na^+ and Mg^{2+} concentrations in the ranges 10–15 m*M* and 0.5–1 m*M* respectively. Omission of either, or both, of these ions from the experiments allows the mitochondria to scavenge Ca^{2+} far more effectively from the medium and maintain a larger Ca^{2+} gradient ($[Ca^{2+}]m > [Ca^{2+}]o$).

The idea that the Ca^{2+} gradient is minimal under appropriate near-physiological conditions also finds support in an experiment on isocitrate oxidation by brown adipose tissue mitochondria (*McCormack* and *Denton* 1980). These mitochondria are uncoupled when prepared and can be recoupled by the addition of bovine serum albumin. This maneuver resulted only in a small (twofold) lowering of the $K_{0.5}$ for extramitochondrial Ca^{2+} for the activation of isocitrate oxidation (*McCormack* and *Denton* 1980). The issue of the concentration gradient across the mitochondrial membrane in vivo is crucial to the theme of this article and will be reconsidered later.

Calcium ions have the effect of decreasing the apparent K_m for isocitrate and 2-oxoglutarate during the state-3 oxidation of these compounds by mitochondria, in analogy with the effect of Ca^{2+} on the isolated enzymes (see Sect. 2). For the effect of Ca^{2+} to be seen, dehydrogenation must contribute largely to the regulation of the overall process of substrate oxidation. This is true, for example, of isocitrate oxidation by adipose tissue mitochondria (*McCormack* and *Denton* 1980); it is untrue for isocitrate oxidation by liver or heart mitochondria, where substrate permeability substantially limits overall substrate oxidation. Table 1 presents a survey of several tissues, with effects of Ca^{2+} on the oxidation of extramitochondrial substrates. It is noted that glutamate oxidation is Ca^{2+} sensitive because of the requirement for 2-oxoglutarate dehydrogenase activity when glutamate is the sole substrate. Under these conditions, oxidation via transaminase, rather than glutamate dehydrogenase, is the expected pathway (*Borst* 1962; *Chappell* 1964; *DeHaan* et al. 1967; *Papa* et al. 1967), and this involves the 2-oxoglutarate dehydrogenase reaction as the initial oxidative step.

Table 1. Sensitivity of mitochondrial substrate oxidation to extramitochondrial Ca^{2+}

Tissue of origin	Degree of respiratory stimulation (% of rate in the absence of Ca^{2+})			Reference
	Isocitrate	2-Oxoglutarate	Glutamate	
Heart	–	180*	121*	This work (see also *Denton* et al. 1980)
Skeletal muscle	–	197*	96	This work
Kidney	80	70	100	This work
Liver	92	92	102	This work
Brain	90	130*	135*	This work
Brown adipose tissue	400	174	–	*McCormack* and *Denton* (1980)

Rates of O_2 uptake were measured with 0.125 m*M* threo-D_s-isocitrate, 0.5 m*M* 2-oxoglutarate, or 2 m*M* L-glutamate as substrate, in the presence of Ca^{2+} (either 8 nmol/mg protein or approx. 10 μM, in the case of brown adipose tissue) or of an excess of EGTA. O_2 uptake was stimulated by ADP, except in the case of brown adipose tissue, where an uncoupling agent was used. A value of 100 indicates an identity of rate in the presence and absence of Ca^{2+}. Values marked with an asterisk show a statistically significant activation of O_2 uptake by Ca^{2+} ($P < 0.05$). The difference in O_2 uptake by adipose tissue mitochondria due to Ca^{2+} is also expected to be significant, in view of the proven significance of the effect of Ca^{2+} on the K_m for isocitrate and for 2-oxoglutarate in that system (*McCormack* and *Denton* 1980). The oxidation of glycerol 3-phosphate by intact mitochondria also shows sensitivity to Ca^{2+}, but this does not involve mitochondrial Ca^{2+} transport; it is discussed in Sect. 2

The sensitivity to Ca^{2+} of the oxidation of glutamate becomes especially clear when studies are carried out in state 4, as opposed to the state-3 studies documented in Table 1. Thus, Fig. 8 presents the results of recent work from this laboratory indicating that the degree of reduction of heart mitochondrial NAD by glutamate in state 4 is markedly dependent on the value of $[Ca^{2+}]o$. It is inferred that the site of the Ca^{2+} sensitivity is the 2-oxoglutarate dehydrogenase and that the heightened effect in state 4 (compare Table 1) reflects the lower concentration of the other activator of 2-oxoglutarate dehydrogenase, ADP, in that state (*Hansford* 1972; *McCormack* and *Denton* 1979). Similar results have been obtained with kidney mitochondria (not shown).

It is noted that insensitivity to Ca^{2+} in Table 1 refers only to the oxidation of added substrate; the inference is that substrate permeation is limiting in these cases, as studies with the isolated NAD-isocitrate and 2-oxoglutarate dehydrogenases have shown Ca^{2+} sensitivity to be ubiquitous in mammalian tissues (*Denton* and *McCormack* 1980). However, the alternative explanation, namely that the $[Ca^{2+}]m$ was never decreased to the range where dehydrogenase modulation is possible, also has to be

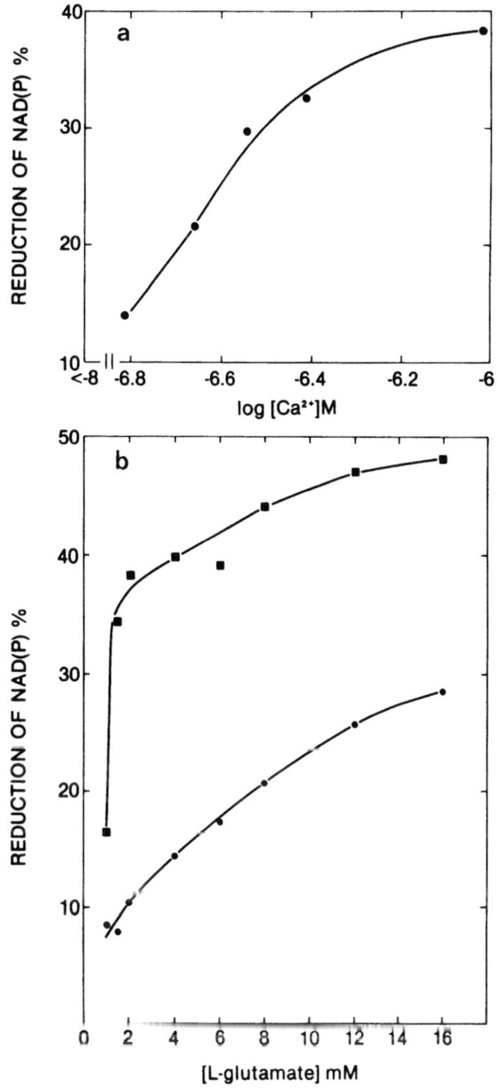

Fig. 8a,b. The activation of glutamate oxidation by $[Ca^{2+}]o$ in heart mitochondria. Dehydrogenase activation was monitored indirectly by measuring the reduction of mitochondrial NAD during state-4 respiration. The activation by Ca^{2+} is construed as evidence for flux limitation by 2-oxoglutarate dehydrogenase during the oxidation of glutamate by heart mitochondria. In **a** the substrate was 2 m*M* L-glutamate and $[Ca^{2+}]o$ was varied as shown by using Ca^{2+}:EGTA buffers. In **b** the glutamate concentration was varied as shown and incubations contained either 2 m*M* EGTA (○) or a Ca^{2+}:EGTA buffer (2 m*M* as EGTA) stabilizing $10^{-6.2}$ *M* free Ca^{2+} (■). All incubations were carried out in a basal medium comprising 0.13 *M* KCl, 20 m*M* K Hepes, 5 m*M* KPi, 10 m*M* NaCl, and 1 m*M* $MgCl_2$

considered, especially for isolated liver mitochondria where a high endogenous content of Ca^{2+} is found after preparation and is experimentally difficult to reduce (see *Borle* 1981 for references).

3.2 Kinetics of Dehydrogenase Activation and Inactivation by Changes in $[Ca^{2+}]o$

The values presented in Figs. 5 and 6 represent steady-state enzyme activities and presumably reflect stable values of $[Ca^{2+}]m$. If one studies the

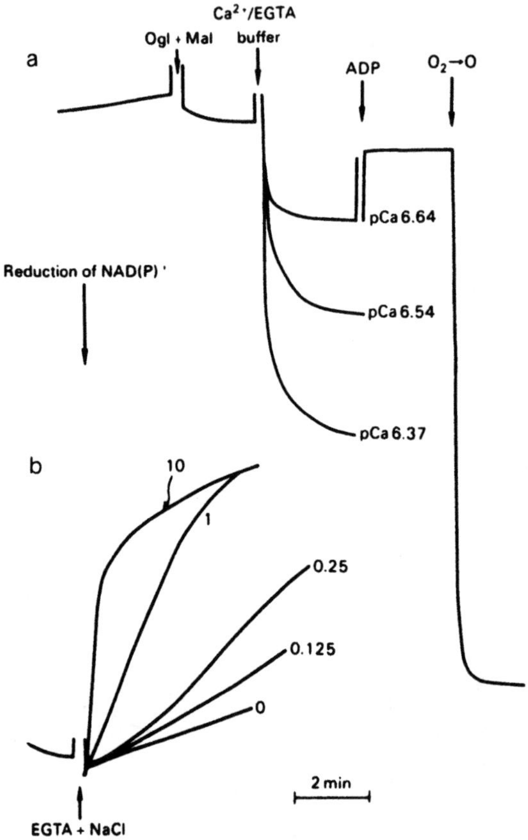

Fig. 9a,b. The kinetics of the activation and inactivation of intramitochondrial 2-oxoglutarate dehydrogenase by changes in $[Ca^{2+}]o$. **a** shows the activation of the oxidation of nonsaturating concentrations of 2-oxoglutarate achieved on adding Ca^{2+}:EGTA buffers of the indicated pCa (-log $[Ca^{2+}]$M) to heart mitochondria which had been previously depleted of endogenous Ca. Dehydrogenase activity was followed by the percentage reduction of mitochondrial NAD in state 4. **b** shows the rate of inactivation of intramitochondrial 2-oxoglutarate dehydrogenase achieved by adding excess EGTA to incubations in a steady state comparable to that seen in **a** at the point of addition of ADP. The rate of inactivation depends upon the concentration of Na^+, which is given in units of m*M* adjacent to the appropriate line. For full experimental details, see the original paper. (*Hansford* and *Castro* 1981)

activity of 2-oxoglutarate dehydrogenase indirectly, via the degree of reduction of components of the respiratory chain, it is possible to monitor enzyme activity continuously and thus to study rates at which activation ba Ca^{2+} and inactivation by Ca^{2+} withdrawal occur. This was done by *Hansford* and *Castro* (1981), who studied the NAD(P)H content of heart mitochondria using fluorimetry. Although both NADH and NADPH contribute to the fluorescence, control experiments showed that changes in fluorescence consequent on 2-oxoglutarate dehydrogenase activation could be attributed almost entirely to changes in NADH content. Figure 9

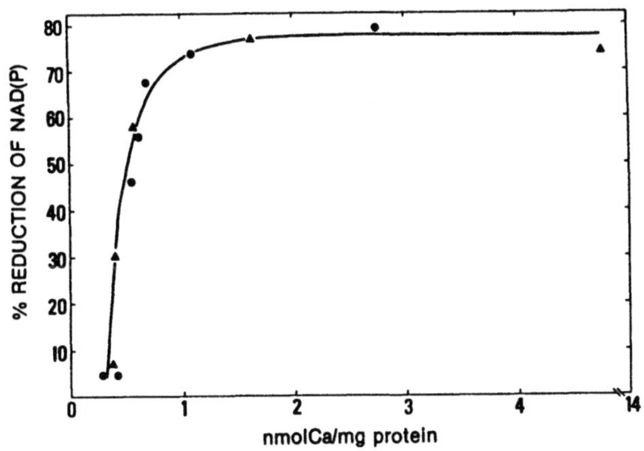

Fig. 10. The dependence of intramitochondrial 2-oxoglutarate dehydrogenase activity upon the total Ca content of rat heart mitochondria. Enzyme activity was assessed by NAD reduction. Mitochondrial Ca content was manipulated by small additions of $CaCl_2$, or of EGTA in the presence of Na^+, and was measured in mitochondrial extracts by atomic absorption spectrophotometry. Full details are given in the original paper. (*Hansford* and *Castro* 1982)

shows that dehydrogenase activation is rapid, even when Ca^{2+} is presented to the mitochondria at (buffered) concentrations of 0.5 μM or less. By contrast, rates of inactivation, associated with the withdrawal of mitochondrial Ca^{2+} by EGTA, are much slower, and depend upon the Na^+ concentration. Rates of activation are Na^+ independent. These experiments demonstrate that the effect of Na^+ in shifting the sensitivity curves of Fig. 5 can indeed be attributed to its effect on Ca^{2+} withdrawal from mitochondria and, by inference, to the operation of the Ca^{2+}/n Na^+ antiporter.

3.3 Dependence of Dehydrogenase Activation on Mitochondrial Total Ca and $[Ca^{2+}]$m

Confirmation that the sensitivity of pyruvate dehydrogenase interconversion and 2-oxoglutarate dehydrogenase activity to Ca^{2+} was similar in the cuvette and in the intramitochondrial milieu came from studies in which enzyme activity was related to mitochondrial Ca content (*Hansford* and *Castro* 1982; *Coll* et al. 1982). As can be seen from Fig. 10, activation of 2-oxoglutarate dehydrogenase occurs at very low levels of total Ca, with 50% activation at approximately 1 nmol Ca/mg protein. Direct measurement of the $[Ca^{2+}]$m corresponding to this value was achieved by an extension of the "null-point" technique previously developed and described by *Williamson* and *Murphy* (1980) and *Murphy* et al. (1980) for the

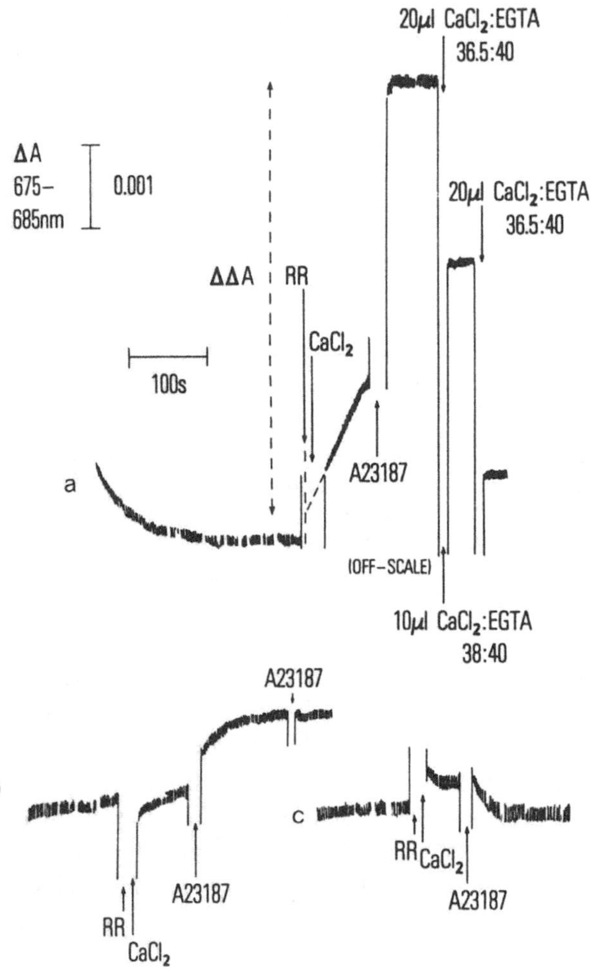

Fig. 11a–c. The protocol for the determination of $[Ca^{2+}]m$ by the "null-point" technique. Experiments used the metallochromic indicator Arsenazo III and the absorbance difference for the wavelength pair 675–685 nm was monitored by using a dual-wavelength spectrophotometer. The change plotted in Fig. 12a (ΔΔA) is the sum of the changes elicited by the addition of ruthenium red (*RR*) and the ionophore *A23187*, using the construction shown, and can be either a net Ca uptake or a net release. In experiment **a** the mitochondria were not depleted of endogenous Ca and the addition of $CaCl_2$ was 70 nmol. In **b** and **c**, the mitochondria were partially depleted by the inclusion of 20 nmol EGTA from the beginning of the experiment. In **b**, the $CaCl_2$ added was 10 nmol; in **c**, it was 40 nmol. The value of $[Ca^{2+}]o$ immediately subsequent to the addition of ruthenium red and $CaCl_2$ was determined in each case by calibration with Ca^{2+}:EGTA buffers of known $[Ca^{2+}]$, as shown in **a**. (*Hansford* and *Castro* 1982)

measurement of free $[Ca^{2+}]$ in both the cytosol and the mitochondria of hepatocytes. Application of this technique to heart mitochondria maintaining contents of total Ca allowing dehydrogenase modulation (Fig. 11) revealed a straight-line dependence of $[Ca^{2+}]m$ on total mitochondrial Ca

Fig. 12a,b. The determination of $[Ca^{2+}]m$ from experimental results of the type shown in Fig. 11, (**a**), and the dependence of $[Ca^{2+}]m$ on the total mitochondrial Ca content (**b**). $[Ca^{2+}]m$ is determined by the null-point technique, corresponding to $\Delta\Delta A = 0$ (**a**), in the absence of a transmembrane pH gradient (see the text). The letters *A* and *B* identify series of experiments with undepleted and partially Ca-depleted mitochondria respectively, and correspond to the labelling in Fig. 11. In **b**, null-point determinations from graphical analysis of the type shown in **a** are plotted against mitochondrial content of total Ca, as determined by atomic absorption spectrophotometry. (*Hansford* and *Castro* 1982)

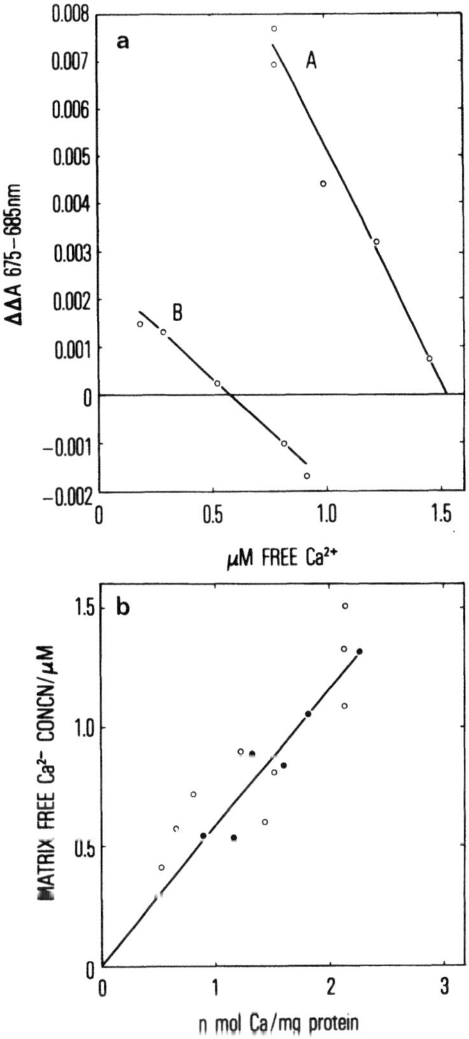

(Fig. 12), with a slope indicating that only 0.06% of the total Ca is free. The exact value of this slope can be challenged, as the null-point technique (*Murphy* et al. 1980) involves the use of the bivalent ionophore A23187 (*Reed* and *Lardy* 1972) to dissipate gradients in Ca^{2+} concentration, and sensitive corrections have to be introduced for the magnitude of the H^+ gradient, as H^+ ions are cotransported by the ionophore (*Reed* and *Lardy* 1972; *Heaton* and *Nicholls* 1976). Small errors in the measurement of the H^+ gradient have a disproportionate effect on the calculated $[Ca^{2+}]$, owing to its dependence on $[H^+]^2$ (see *Coll* et al. 1982); moreover, the ionophore also transports Mg^{2+} ions, such that a preexisting gradient of Mg^{2+} ions prior to the addition of the ionophore would be expected to lead to a flux of Ca^{2+} ions on the addition of A23187, when there is no preexisting gradient of Ca^{2+} ion concentration. These cautions notwithstanding, these experiments and similar ones performed by *Coll* et al. (1982) provide encouraging support for the idea that, given total Ca loads in the region allowing dehydrogenase modulation (0–5 nmol/mg protein), the parameter $[Ca^{2+}]m$ is in the μM range and not two orders of magnitude higher as surmised by *Nicholls* (1978). Moreover, these results describe the sensitivity of pyruvate dehydrogenase interconversion and 2-oxoglutarate dehydrogenase activity to $[Ca^{2+}]$, when these enzyme systems are present in their intramitochondrial locale. Comparison of Figs. 10 and 12b suggests a $K_{0.5}$ for Ca^{2+} of approximately 0.6 μM for intramitochondrial 2-oxoglutarate dehydrogenase, a value similar to that found for the purified enzyme (*McCormack* and *Denton* 1979). The quotient "$K_{0.5}$ for activation of 2-oxoglutarate dehydrogenase in the mitochondrial matrix" (as just derived): "$K_{0.5}$ for activation of 2-oxoglutarate dehydrogenase in intact mitochondria" (0.3 μM from Fig. 6) then gives a mitochondrial membrane concentration gradient for Ca^{2+} of approximately 2 (in > out). As mentioned above, this is larger if Mg^{2+} or Na^{+} is omitted from the studies with intact mitochondria (see Figs. 5, 6).

It is noted that the relationship between total mitochondrial Ca and $[Ca^{2+}]m$ is similar in liver and heart mitochondria (*Coll* et al. 1982) but that the sensitivity of 2-oxoglutarate dehydrogenase to $[Ca^{2+}]m$ has been investigated only in heart mitochondria (*Hansford* and *Castro* 1982; *Coll* et al. 1982).

4 The Calcium Content of Mitochondria in Situ

The thesis of dehydrogenase control by Ca^{2+} would be strengthened by the demonstration that mitochondria in situ actually contain total Ca in the region of 0–3 nmol/mg mitochondrial protein, this being the range shown to allow dehydrogenase modulation in heart mitochondria (Fig. 10). This section examines the evidence in favour of this proposal.

4.1 Ca Content of Muscle Mitochondria

A strong claim can be made that the Ca content of muscle mitochondria in situ is below 3 nmol/mg mitochondrial protein, provided that the sarcolemma has not been damaged during tissue preparation and fixation. This statement is based upon electron-probe X-ray microanalysis results from frozen and unfixed samples of vascular smooth muscle (*Somlyo* et al. 1979, 1981) and cardiac muscle (*Wendt-Gallitelli* and *Jacob* 1982).

A great strength of this method is that it can evaluate the integrity of individual cells, with the finding that mitochondria containing granules rich in Ca and P seem to be associated only with cells containing lowered cytosolic K^+:Na^+ ratios, and in which the permeability properties of the plasma membrane have presumably been compromised (*Gupta* and *Hall* 1978; *Somlyo* et al. 1979). In view of the magnitude of the inwardly directed Ca^{2+} concentration gradient across the plasma membrane of most cells (see *Borle* 1981, and *Rasmussen* and *Waisman* 1983 for reviews) and of the potential activity of the Ca^{2+} uniport of most mitochindria at Ca^{2+} concentrations greater than 1 or 2 μM (see *Fiskum* and *Lehninger* 1982; *Nicholls* and *Åkerman* 1982, for references), it is not surprising that perturbations in plasma membrane permeability lead to massive accumulations of Ca within mitochondria.

A relative weakness of the X-ray microanalysis method is its limited sensitivity; however, 1 m*M* Ca can be detected (*Gupta* and *Hall* 1978) and this would correspond to appoximately 1 nmol total Ca/mg mitochondrial protein, making a reasonable assumption of matrix volume. This being so, amounts of mitochondrial Ca which would activate pyruvate dehydrogenase interconversion and 2-oxoglutarate dehydrogenase (Fig. 10) by 50% should be detectable by this technique. Significantly, the X-ray microanalysis method reveals that the magnitude of the ratio of specific mitochondrial Ca content to specific cytosolic Ca content is preserved near unity when smooth muscle is subjected to a lengthy depolarization (*Somlyo* et al. 1979). Absolute values increase and the increase in the mitochondrial compartment would be consistent with the activation of catabolic metabolism, which is the theme of this article.

A low in situ Ca content of cardiac muscle mitochondria also emerges from conventional homogenization and fractionation experiments, provided that care is taken to minimize Ca^{2+} accumulation during tissue disruption (*Henry* et al. 1977; *Hunter* et al. 1980). This has been done either with ruthenium red, which was indicated by controls to greatly diminish, but not abolish, uptake (*Henry* et al. 1977), or with chelating agents. In the latter case, a prior perfusion of the vasculature with a medium containing EGTA materially lowers the Ca content of the mitochondrial fraction relative to a preparation in which EGTA is included in the medium but in which there is no prior perfusion (*Hunter* et al. 1980). Clearly, these protocols can be criticized on the grounds that the uptake of ambient Ca^{2+} is ruled out, or at least curtailed, during the preparation of the mitochondrial fraction, whereas egress is not inhibited, except by the use of low temperature (0°). However, control experiments indicate that release of previously accumulated Ca^{2+} was minimal in these experiments (*Henry* et al. 1977; *Hunter* et al. 1980). Nevertheless, values of total Ca content of 0.25 nmol/mg mitochondrial protein (using a figure of 40 mg mitochondrial protein per g wet wt of heart) obtained after EGTA perfusion (*Hunter* et al. 1980) should probably be regarded as minimal estimates of in situ content.

The technique of investigating intracellular Ca compartments by the kinetic analysis of ^{45}Ca uptake and, particularly, release (see *Borle* 1981 for a review) has been applied to cardiac muscle but has not provided unequivocal information on the size of the mitochondrial pool. In the isolated perfused heart, the kinetic compartment with the lowest exchange rate is identified with the mitochondria, with some possible contribution from the sarcoplasmic reticulum (*Barritt* and *Spiel* 1981). This assignment was based upon the rate of ^{45}Ca loss from subcellular fractions obtained by homogenization and conventional separation techniques in the presence of EGTA. The pool thus identified is increased in glucagon-treated hearts, but its absolute size is not clear. In experiments with cultured cardiac myocytes, a pool that is slowly labelled with ^{45}Ca was identified as being mitochondrial on the grounds that uptake was sensitive to the inhibition of respiration by warfarin and was potentiated by phosphate (*Langer* and *Nudd* 1980). Phosphate greatly increased the size of this Ca pool, compared to media lacking ions capable of entry into the mitochondria as the undissociated free acid or by proton symport. In the presence of phosphate, the mitochondrial pool exceeded 3 mmol/kg wet wt – corresponding to approximately 80 nmol Ca/mg mitochondrial protein (assuming 40 mg mitochondrial protein/g wet wt). This figure is much larger than those presented earlier in this section, obtained using other techniques. One reason for this might be the use of a supraphysiological concentration of phosphate (10 m*M*) in this study; an alternative tentative explanation might be an increased sarcolemmal Ca^{2+} permeability in the cultured cells relative to the tissue of origin.

4.2 Ca Content of Liver and Kidney Mitochondria

Although there is a wide range of experimental values for the Ca content of mitochondria in situ in liver and kidney cells, there are a number of observations suggesting that this content is in the range permitting dehydrogenase modulation by changes in $[Ca^{2+}]m$. The experimental approaches used include (a) cell fractionation using either conventional or rapid centrifugation, and in the presence or absence of inhibitors of transport; (b) analysis of the kinetics of ^{45}Ca uptake and/or release; and (c) identification of mitochondrial Ca as the fraction of cellular Ca released by uncoupling agents. The field has recently been reviewed in this series by *Åkerman* and *Nicholls* (1983), with particular emphasis on the effects of α_1-agonists, and the brief statement that follows will attempt to avoid redundancy.

Fractionation Studies. Studies in which total mitochondrial Ca content was measured after perfusion of liver, followed by the isolation of mitochondria in the presence of EGTA, have yielded values in the range 4–5 nmol of Ca/mg mitochondrial protein (*Hughes* and *Barritt* 1978; *Blackmore* et al. 1979a). One of these studies (*Blackmore* et al. 1979a) additionally included 1 m*M* $LaCl_3$, shown by *Crompton* et al. (1979) to be an effective inhibitor of both uptake and release pathways. Perfusion of the liver with a medium containing a reduced Ca^{2+} concentration (50 μM) resulted in a lowering of mitochondrial Ca content to 1.9 nmol/mg mitochondrial protein, thereby reproducing the results of an earlier study which used a perfusion medium containing a low Ca^{2+} concentration (*Blackmore* et al. 1979b).

Omission of an inhibitor of the mitochondrial Ca^{2+} release pathway raises the possibility that net egress of Ca^{2+} may occur during isolation of mitochondria, though in fact the results of *Blackmore* et al. (1979a), involving experiments containing $LaCl_3$, and those of *Hughes* and *Barritt* (1978), involving no inhibitor of egress, were similar. Additionally, control experiments directed at this question in fact showed very little loss of accumulated Ca during a conventional isolation of mitochondria from liver (*Claret-Berthon* et al. 1977). In view of the extremely high potential activity of the Ca^{2+} uniport in liver mitochondria (see, e.g., *Vinogradov* and *Scarpa* 1973; *Bragadin* et al. 1979), it seems quite likely that the net artifact is in the opposite direction, and that Ca^{2+} uptake occurs during tissue fragmentation. This would happen unless access of the homogenization medium to the mitochondria is as rapid as is access of the blood or perfusion fluid present within the tissue; this would seem unlikely. Some such mechanism would seem to underlie the differences seen on introducing a perfusion of heart muscle with medium containing EGTA immediately prior to homogenization (*Hunter* et al. 1980).

Similar fractionation studies with isolated liver cells and slices generally reveal higher mitochondrial Ca contents (*Foden* and *Randle* 1978; *Van Rossum* et al. 1976; *Murphy* et al. 1980; *Joseph* et al. 1983). Despite this near-consensus, it seems, however, premature to conclude that the high Ca content of tissue slices and hepatocytes reflects the situation in the animal. The problem of heterogeneity is a real one: mitochondria in a small fraction of damaged cells exposed to the 1–2 m*M* Ca^{2+} of the perfusion fluid will weight the average of the overall mitochondrial population. Such overloaded mitochondria in damaged cells have been seen directly by electron-probe analysis (*Somlyo* et al. 1981) of vascular smooth muscle and derived mitochondrial preparations.

Kinetic Studies. The measurement of in situ mitochondrial Ca content by kinetic analysis of ^{45}Ca desaturation curves has been recently reviewed by

Borle (1981) in this series. When applied to perfused liver, the technique has provided evidence for an exchangeable mitochondrial Ca content of 0.12 μmol/g wet wt (*Claret-Berthon* et al. 1977), or approximately 2.4 nmol/mg mitochondrial protein on the assumption of 50 mg mitochondrial protein/g liver (*Carafoli* 1967). This is of the same order of magnitude as the range of Ca loads allowing dehydrogenase control in heart mitochondria (*Hansford* and *Castro* 1982; *Coll* et al. 1982). As the fraction of total Ca existing as free Ca^{2+} in the matrix is quite similar for liver and heart mitochondria (*Coll* et al. 1982), this would seem to be consistent with the operation of such dehydrogenase control in the liver. The argument is complicated, however, by the fact that the content of 2.4 nmol Ca/mg mitochondrial protein (*Claret-Berthon* et al. 1977) is the fraction which can be exchanged, not the total.

Similar experiments with isolated hepatocytes have also indicated a very low value for exchangeable mitochondrial Ca (*Poggioli* et al. 1980). Here, again, the exchangeable fraction may be significantly less than the total, as judged from the experiments by *Foden* and *Randle* (1978), who also used isolated hepatocytes.

Application of these techniques to isolated kidney cells gives control values of total exchangeable mitochondrial Ca of 2.4–3 nmol/mg cellular protein (*Studer* and *Borle* 1979). On the basis of mitochondrial protein, these values would appear to exceed the control range defined for (heart) dehydrogenases.

Uncoupling Agent Studies. The Ca content of mitochondria has also been estimated on the basis of the fraction of cellular Ca released on treatment with uncoupling agents, i.e., proton ionophores (*Bygrave* 1977). When applied to suspensions of hepatocytes either prepared conventionally or with a Ca-loading step, this technique established the mitochondria as the major Ca sink within the cell, with a content of Ca equal to 65%–80% of the cellular total (*Murphy* et al. 1980; *Joseph* et al. 1983). It is hard to accept that release by an uncoupling agent is a rigorous measure of mitochondrial Ca, as the hydrolysis of cytoplasmic ATP by uncoupled mitochondria could surely compromise Ca^{2+} transport by other, ATP-driven, membrane systems. However, the technique gains credibility in the study of *Murphy* et al. (1980) in that an independent assessment of mitochondrial Ca on the basis of cell fractionation provided similar answers. The cell disruption and fractionation technique should have given less scope for an artifactual redistribution of Ca than conventional cell fractionation, as it used a combination of digitonin treatment and shearing forces for very rapid removal of the plasma membrane, followed by rapid centrifugation. Ruthenium red was present to inhibit uptake. These studies indicated a mitochondrial content of total Ca of approximately 10 nmol/mg mitochondrial protein for cells not deliberately loaded with Ca (*Murphy* et al. 1980). A later paper by the same group (*Joseph* et al. 1983) reported a somewhat higher figure, a minimum of 16 nmol/mg mitochondrial protein for such cells, again based upon the criterion of releasability by uncoupling agents.

The application of a similar approach to isolated kidney tubules also identified mitochondria as the site of 80% of total cellular Ca (*Murphy* and *Mandel* 1982). In unloaded tubules, this corresponded to approximately 10 nmol Ca/mg cellular protein.

If truly characteristic of the Ca content of liver and kidney mitochondria in the living animal, these values of uncoupling agent–releasable Ca would rule out dehydrogenase control by Ca^{2+} in these tissues, and indeed that is the point of view taken by *Joseph* et al. (1983) and by *Williamson* and *Cooper* (1980). However, there is more compelling evidence in favor of dehydrogenase control based on other criteria (see below; Sect. 5).

4.3 Ca Content of Nervous Tissue Mitochondria

The in situ Ca content of nervous tissue mitochondria has been approached with the same armory of techniques described above for liver and kidney, with the general finding that this content is slightly in excess of the range defined as allowing dehydrogenase control in heart mitochondria (*Hansford* and *Castro* 1981, 1982; *Coll* et al. 1982). However, the same reservations apply. Experiments with isolated synaptosomes indicate a mitochondrial Ca content of 2.2 nmol/mg synaptosomal protein when the mitochondrial pool is measured by disruption of the plasma membrane by digitonin and shearing forces followed by rapid microcentrifugation (*Scott* et al. 1980). Depolarization of the plasma membrane results in raised total synaptosomal Ca contents, with an increased proportion being found in the mitochondrial fraction (*Scott* et al. 1980; *Åkerman* and *Nicholls* 1981a). Although Ca^{2+} uptake by mitochondria in damaged synaptosomes was thought not to prejudice these valves, on the grounds that ruthenium red had no effect (*Åkerman* and *Nicholls* 1981a), the possibility remains that plasma membrane permeability properties or transport activity may be changed from that in vivo. Indeed, a preincubation phase in the absence of Ca^{2+} was found to markedly decrease the total content of Ca when synaptosomes subsequently incubated in media containing a physiological Ca^{2+} concentration were compared with synaptosomes not subjected to this prior Ca^{2+}-free incubation (*Scott* et al. 1980). Thus, one should probably be cautious in relating these absolute values of synaptosomal mitochondrial Ca content to those obtaining in situ, and then inferring whether dehydrogenase modulation by Ca^{2+} is feasible or not. However, these studies show clearly that the increase in the value $[Ca^{2+}]c$ found on depolarization of nerves (*Katz* and *Miledi* 1967a,b; *Blaustein* 1971; *Llinás* et al. 1972) is associated with a major increase in the Ca content of the mitochondrial compartment (*Scott* et al. 1980; *Åkerman* and *Nicholls* 1981a). A similar conclusion emerges from ^{45}Ca desaturation studies of rat anterior pituitary slices (*Moriarty* 1980), in which depolarization by high concentrations of extracellular K^+ resulted in the enlargement of a slowly exchanging pool tentatively identified with mitochondria and/or microsomes. The magnitude of this pool was around 4 nmol/mg cell protein in the control incubations.

Finally, electron-probe X-ray microanalysis has visualized an increase in mitochondrial Ca content in presynaptic nerve endings when cervical sympathetic nerves were stimulated electrically (*Parducz* and *Joó* 1976). Absolute Ca contents are not available from this report, and might not be reliable, owing to the use of chemical fixation involving high Ca^{2+} concentrations and high temperatures. Nevertheless, the difference in mitochondrial Ca content seen on electrical stimulation remains provocative.

It is concluded that too little information is available on in situ mitochondrial Ca contents and on the degree of intramitochondrial buffering of Ca^{2+} in nervous tissue for any comment to be made on whether dehydrogenase modulation by Ca^{2+} ions is feasible or not, on the basis of measured mitochondrial Ca content and calculated intramitochondrial free Ca^{2+} concentration. However, other arguments make such control very credible, as developed below. In addition, the increase of mitochondrial Ca content in response to depolarization (*Scott* et al. 1980; *Åkerman* and *Nicholls* 1981a; *Moriarty* 1980; *Parducz* and *Joó* 1976) may then provide the signal for the increase in catabolic flux necessary to sustain cellular adenine nucleotide phosphate potential (see Sect. 8).

4.4 Consequences for the Model of Dehydrogenase Regulation

Section 4.1 has provided reasonable grounds for the statement that the content of mitochondrial Ca (i.e., total, bound plus free Ca) in situ in cardiac and vascular smooth muscle corresponds with the range of mitochondrial Ca contents shown in in vitro studies to give modulation of dehydrogenase activity in heart mitochondria (*Hansford* and *Castro* 1981, 1982; *Coll* et al. 1982). Extension of this conclusion to other tissues is quite premature, for two reasons.

One reason is that the relationship between total mitochondrial Ca content and matrix free Ca^{2+} concentration, i.e. the characterization of intramitochondrial Ca^{2+} buffering, has only been approached in two tissues, namely heart (*Hansford* and *Castro* 1982; *Coll* et al. 1982) and liver (*Coll* et al. 1982). There is no reason to assume that Ca^{2+} buffering will be the same in disparate tissues, though in fact results from liver and heart correspond quite closely. Equally, in vitro and in vivo buffering may be different for mitochondria from any given tissue, depending on, for instance, the phosphate concentration used in in vitro studies. Finally, the existence of more than one pool of intramitochondrial Ca, as described for heart mitochondria by *Barritt* and *Lamont* (1982), would complicate the relationship between total Ca and $[Ca^{2+}]m$ in the vicinity of the dehydrogenases.

The second reason is that there is great uncertainty as to what the in situ Ca content of mitochondria in liver, kidney, and nervous tissue really is. This is evidenced, for instance, by the range of synaptosomal Ca contents, from less than 1 to 30 nmol/mg protein, found in studies using media containing physiological Ca^{2+} concentrations (*Blaustein* 1975; *Blaustein* and *Ector* 1976; *Jansson* et al. 1977; *Wonnacott* et al. 1978). In general, Ca contents appear to be too high to allow dehydrogenase modulation by Ca^{2+} in tissues other than cardiac and vascular smooth muscle. However, as outlined above, methodological artificats tend to increase mitochondrial Ca contents. It seems that very rapid freezing and electron-probe X-ray microanalysis provide the most reliable measurements of tissue Ca content (*Somlyo* et al. 1978, 1979, 1981), and application of these techniques to a wider range of tissues should be most worthwhile.

5 Values of $[Ca^{2+}]c$. Relevance to the Model of Dehydrogenase Regulation

Far stronger support for the model of dehydrogenase regulation by Ca^{2+} emerges from a consideration of values of $[Ca^{2+}]c$. It was established in Sect. 3.1 that exposure of intact, nonpoisoned, mitochondria from heart, skeletal muscle, kidney, adipose tissue, and brain, incubated under physiologically appropriate ionic conditions (120 m*M* K^+, 10 m*M* Na^+, 1 m*M* Mg^{2+}, 5 m*M* Pi), to media containing various free Ca^{2+} concentrations gave 50% activation of pyruvate dehydrogenase interconversion or of 2-oxoglutarate dehydrogenase at values of $[Ca^{2+}]o$ of 0.25–0.35 μM. If values of $[Ca^{2+}]c$ fall within this range, then modulation of intramitochondrial dehydrogenases by changes in $[Ca^{2+}]$ becomes eminently reasonable. This section presents evidence that this is in fact so.

5.1 Heart Muscle

Until recently, the only direct measurements of $[Ca^{2+}]c$ were confined to the giant nerve and muscle fibers of lower animals (see reviews by *Borle* 1981; *Blinks* et al. 1982). Recently, however, both injection of the luminescent protein aequorin (*Allen* and *Blinks* 1978) and impalement with Ca^{2+}-specific microelectrodes (*Lee* et al. 1980; *Marban* et al. 1980) have provided values for $[Ca^{2+}]c$ in vertebrate heart cells. The electrode measurements indicate that the resting $[Ca^{2+}]c$ of cardiac muscle is 0.12 μM (*Lee* et al. 1980) or 0.26 μM (*Marban* et al. 1980), with the small difference between these published values being entirely due to different assumptions of the stability constant for the Ca:EGTA complex (*Marban* et al. 1980). On contraction of the muscle preparations, $[Ca^{2+}]c$ was found to rise to 2 μM in the electrically stimulated frog preparation with aequorin as indicator (*Allen* and *Blinks* 1978), and as high as 10 μM during contractures of ferret heart induced by Na^+ replacement and by caffeine (*Marban* et al. 1980). The latter study used Ca^{2+} microelectrodes. Lesser tension development, caused solely by replacement of extracellular Na^+, was associated with a lesser rise in $[Ca^{2+}]c$ (*Lee* et al. 1980; *Marban* et al. 1980).

When these values are compared with the sensitivity of the intramitochondrial enzymes to $[Ca^{2+}]o$ (Figs. 5, 6), it becomes apparent that enzyme activity would be near minimal in the resting state and maximal at a value of $[Ca^{2+}]c$ lower than that generated during contraction. The effective stability constant for Ca:EGTA used by *Hansford* (1981) and *Hansford* and *Castro* (1981) was $10^{7.08}$ at pH 7.2 and $10^{6.68}$ at pH 7, and thus falls between the constants used by *Lee* et al. (1980) and *Marban* et al. (1980) in their calculations, which were $10^{6.76}$ and $10^{6.42}$ respectively at pH 7. Probably, the meaningful parameter is the time-averaged value of $[Ca^{2+}]c$ of cardiac muscle, as there is great doubt as to the ability of the mitochondrial Ca^{2+} transport systems to follow beat-to-beat fluctuations in $[Ca^{2+}]c$. This follows from the demonstrated low activity of the Ca^{2+} uniporter at prevailing cytosolic concentrations of Ca^{2+} and Mg^{2+} (*Scarpa* and *Graziotti* 1973) and from the relatively low V_{max} of the Ca^{2+}/n Na^+ antiport pathway (*Hansford* and *Castro* 1982). In support of this view, *Robertson* et al. (1982) have calculated that not more than 3% of the Ca^{2+} exchanged with the myofibrils during a beat will be accumulated by the mitochondria, on the basis of their Ca^{2+}-uptake results with pig heart mitochondria and the transient kinetics of aequorin luminescence in stimulated frog heart cells (*Allen* and *Blinks* 1978). As the authors acknowledged, such an estimate is apparently a maximum in that they ignored the presence of the electroneutral antiporter (*Robertson* et al. 1982), which will result in a net accumulation of mitochondrial Ca during systole which is less than the computed flux through the uniporter. It seems that $[Ca^{2+}]m$ will follow changes in $[Ca^{2+}]c$ in cardiac muscle, but with a large degree of damping. Just how great this degree of damping is awaits measurements of $[Ca^{2+}]c$ transients in mammalian cardiac cells, as the amphibian heart handles Ca^{2+} differently (*Morad* and *Goldman* 1973), and more sophisticated modelling. However, it is clear that more frequent contractions or an increased $[Ca^{2+}]c$ transient during systole, as in the inotropic response to catecholamines (*Allen* and *Blinks* 1978), would be expected to raise the time-average value of $[Ca^{2+}]c$ and thus of $[Ca^{2+}]m$ and thereby result in increased activity of the Ca^{2+}-sensitive intramitochondrial dehydrogenases. Evidence that this is in fact so is presented in Sect. 7.1.

It is noted that in an excitable tissue which experiences marked transients in $[Ca^{2+}]c$, the periods of elevated $[Ca^{2+}]c$, for instance, systole in the heart, will have a weighted effect on $[Ca^{2+}]m$ owing to the great potential activity (i.e., V_{max}) and sigmoidal concentration dependence of the Ca^{2+} uniporter (*Scarpa* and *Graziotti* 1973). Thus, it seems there may be a "memory" inherent in the mitochondrial uptake systems, with the consequence that the time-average concentration gradient $[Ca^{2+}]m/[Ca^{2+}]c$ may be greater in a tissue with a periodic variation in $[Ca^{2+}]c$ than in a tissue in which this is rather constant in the short term.

Prolonged elevation of $[Ca^{2+}]c$ as during contracture (*Marban* et al. 1980) would be expected to give such a massive net flux of Ca^{2+} into mitochondria that an early activation of Ca^{2+}-sensitive dehydrogenases might be succeeded by a respiratory decline, owing to mitochondrial swelling and loss of cofactors from the matrix space. Such mitochondrial pathology is in fact evident when heart mitochondria are incubated for a matter of minutes in media containing buffered Ca^{2+} at 10^{-6} *M* or higher concentrations (*Hansford* and *Cohen* 1978; *Robertson* et al. 1982).

5.2 Nervous Tissue

In the nervous system, it is well accepted that an increase in $[Ca^{2+}]c$ in presynaptic nerve endings forms part of the mechanism of transmitter substance release (*Katz* and *Miledi* 1967a,b; *Llinás* et al. 1972; *Blaustein* 1975; *Llinás* and *Heuser* 1977). Although an increase in $[Ca^{2+}]c$ has been shown on depolarization, using the bioluminescent indicator aequorin (see, e.g., *Llinás* et al. 1972), there seems to be no quantitative measure of $[Ca^{2+}]c$ upon depolarization. However, *Åkerman* and *Nicholls* (1981b) have approached this question indirectly by monitoring the increase in mitochondrial respiration when the plasma membrane of synaptosomes (pinched-off presynaptic nerve endings) is depolarized. Calibration with the respiratory response to Ca^{2+} of isolated mitochondria suggested that the synaptosomal $[Ca^{2+}]c$ reaches $10^{-5}-10^{-6}$ *M* under conditions of external Ca^{2+} concentration and bivalent cation ionophore availability which allowed the release of the transmitter substance norepinephrine. An increase in respiration by intrasynaptomal mitochondria in response to Ca^{2+} (*Åkerman* and *Nicholls* 1981b) betokens an increase in $[Ca^{2+}]c$ to values in excess of 1 *μM* and uniporter flux sufficient to diminish the mitochondrial proton-motive force. If continued, this situation would lead to massive Ca^{2+} accumulation by the mitochondria and, eventually, damage to these organelles. That it evidently does not continue, in vivo, must reflect the transport of Ca^{2+} out of the nerve by plasma membrane systems. The competition between such systems and mitochondrial Ca^{2+} transport in the maintenance of $[Ca^{2+}]c$ is discussed in Sect. 6.

In the context of dehydrogenase regulation, it is apparent from results with mitochondria isolated from central nervous tissue (Fig. 5a) that $[Ca^{2+}]c$ values in the range $10^{-5}-10^{-6}$ *M* (*Åkerman* and *Nicholls* 1981b) will more than suffice to stimulate intramitochondrial dehydrogenases during depolarization of presynaptic nerve endings. Equally, resting values of $[Ca^{2+}]c$, at least as measured in squid axons (*Baker* 1972; *DiPolo* et al. 1976), are low enough to suppress intramitochondrial dehydrogenase activity. Thus, dehydrogenase control by Ca^{2+} is eminently plausible in nervous tissue. The degree of activation of these enzymes achieved under physiological circumstances will depend upon the kinetics of changes in $[Ca^{2+}]c$ and on the degree of damping imposed by the mitochondrial Ca^{2+}-transport pathways, in analogy with the arguments concerning heart (Sect. 5.1). Further discussion at this time would be speculative.

5.3 Liver and Kidney

Recently, direct measurements of $[Ca^{2+}]c$ have been extended to include small mammalian cells. This was first accomplished by the development of the null-point technique (*Williamson* and *Murphy* 1980; *Murphy* et al. 1980), already described briefly in the context of the measurement of $[Ca^{2+}]m$. Applied to hepatocytes this gave a value for $[Ca^{2+}]c$ of 0.19 *μM*, rising to 0.46 *μM* in response to phenylephrine (*Murphy* et al. 1980). Quite impressively, the content of phosphorylase *a*, which can be used as an indirect indicator of $[Ca^{2+}]c$ owing to its established Ca^{2+} sensitivity (*Ozawa* 1972), was found to parallel the changes in $[Ca^{2+}]c$ almost exactly, both as a function of time and of the concentration of α_1-agonist. Applied to isolated mammalian kidney tubules, the null-point technique revealed a somewhat higher $[Ca^{2+}]c$, namely 0.45 *μM* (*Murphy* and *Mandel* 1982).

These values may be compared with the results from two other techniques, both of which have been described within the last year or so. The fluorescent Ca^{2+}–chelating agent Quin 2, developed by *Tsien* et al. (1982), is accumulated by cells containing esterase activity when presented as the acetoxymethyl ester. It shows an increase in fluorescence in response to Ca^{2+}, with high specificity, and appears not to be sensitive to binding artifacts and to be excluded from the mitochondrial compartment. Quin 2 thus presents a relatively convenient way of monitoring steady-state values of $[Ca^{2+}]c$; it necessarily buffers transient changes owing to its being an effective chelating agent, with a dissociation constant of the Quin 2:Ca^{2+} complex of 115 n*M* (*Tsien* et al. 1982). Studies of the fluorescence of Quin 2–loaded hepatocytes (*Charest* et al. 1983) revealed a resting $[Ca^{2+}]c$ of approximately 0.2 μM, rising to approximately 0.6 μM in response to epinephrine or phenylephrine. The agreement with the results of the null-point experiments of *Murphy* et al. (1980) is really quite impressive. This mobilization of Ca^{2+} in response to α_1-agonists will be discussed further below (Sect. 7.3).

Finally, there is one recent instance in which aequorin has been used as a Ca^{2+} indicator in small mammalian cells. This involved a hypotonic shock treatment of the cells and gave a result of 57 n*M* for $[Ca^{2+}]c$ in cultured monkey kidney cells (*Borle* and *Snowdowne* 1982). This value is somewhat lower than that obtained with the null-point technique using freshly isolated kidney tubules (*Murphy* and *Mandel* 1982).

With regard to dehydrogenase control, unpublished work from this laboratory shows that the activity of 2-oxoglutarate dehydrogenase in kidney mitochondria displays much the same response to $[Ca^{2+}]o$ as does the enzyme in heart mitochondria (Fig. 6). Thus, 57 n*M* Ca^{2+} (*Borle* and *Snowdowne* 1982) would give minimal activation, whereas 0.45 μM (*Murphy* and *Mandel* 1982) would give somewhat more than 50% activation of the intramitochondrial enzyme. Clearly, within the limits of precision of the available data, $[Ca^{2+}]c$ of kidney cells lies within the range allowing dehydrogenase modulation. If the increase in $[Ca^{2+}]c$ shown in liver on activation with α_1-agonists (*Murphy* et al. 1980; *Charest* et al. 1983) also occurs in kidney, then a rise in $[Ca^{2+}]m$ sufficient to activate 2-oxoglutarate dehydrogenase would be expected to occur.

Such interpretations must be more guarded for liver. This arises because of the failure so far to establish Ca^{2+} sensitivity of substrate oxidation by intact liver mitochondria (Table 1), despite the proven Ca^{2+} sensitivity of pyruvate dehydrogenase phosphate phosphatase, NAD-isocitrate dehydrogenase, and 2-oxoglutarate dehydrogenase purified or partially purified from that tissue (*Denton* and *McCormack* 1980). If this failure reflects only rate limitation of exogenous substrate oxidation by substrate permeability, then Ca^{2+} may still be a physiologically important modulator of the oxidation of isocitrate and 2-oxoglutarate generated within the mitochondria. This needs to be studied, as does the response of the PDH_A

content of liver mitochondria to changes in [Ca^{2+}]o. If, on the other hand, the lack of response of substrate oxidation to Ca^{2+} in in vitro experiments (Table 1) arises through a mitochondrial Ca content which is continuously saturating the dehydrogenase systems, the question arises of whether this is an artifact of preparation or a true representation of in situ Ca content of liver mitochondria. This question was addressed in Sect. 4.2.

6 Does Mitochondrial Ca^{2+} Transport Control [Ca^{2+}]m or [Ca^{2+}]c?

The view has arisen that the mitochondrial Ca^{2+} transport processes serve to buffer or maintain particular values of [Ca^{2+}]c (for reviews see *Borle* 1981; *Nicholls* and *Åkerman* 1982; *Fiskum* and *Lehninger* 1982; *Åkerman* and *Nicholls* 1983). The maintenance of [Ca^{2+}]c is not the primary concern of this article. However, discussion of this point of view is unavoidable, as an effective role of mitochondria in buffering the [Ca^{2+}] of the cytosol would preclude meaningful control of intramitochondrial dehydrogenases by Ca^{2+}. The reason that the control of the central intramitochondrial catabolic pathways by Ca^{2+} and the buffering of [Ca^{2+}]c are not compatible lies in the nature of the mitochondrial Ca^{2+} release pathway, which is discussed below.

6.1 The Kinetics of Ca^{2+} Efflux in Heart and Liver Mitochondria and the Relationship Between [Ca^{2+}]c and [Ca^{2+}]m

When the egress of Ca^{2+} was studied in suspensions of heart mitochondria with a low total Ca content, a linear relationship between 1/V and 1/S was established, where V is the velocity of Ca^{2+} efflux and S is the total mitochondrial Ca content as determined by atomic absorption spectrophotometry (*Hansford* and *Castro* 1982). This relationship is presented as Fig. 13. Comparison with the results of null-point studies of heart [Ca^{2+}]m as a function of total mitochondrial Ca (Fig. 12b) allows conversion of the S values in Fig. 13 to concentrations of free Ca^{2+}. Thus, the $S_{0.5}$ value of 5.5 nmol Ca/mg mitochondrial protein (Fig. 13) is equivalent to a value of [Ca^{2+}]m of 3.2 μM. This compares with an $S_{0.5}$ value of 5.7 μM determined by *Coll* et al. (1982) using closely similar methodology and reasoning. Both values are somewhat lower than the $S_{0.5}$ value for external Ca^{2+} determined by *Crompton* et al. (1977) in studies of the Ca^{2+}-stimulated release of $^{45}Ca^{2+}$ from heart mitochondria (13 μM). The discrepancy might reflect systematic error in the estimation of the "activity coefficient" of intramitochondrial Ca in the recent work (*Hansford* and *Castro* 1982;

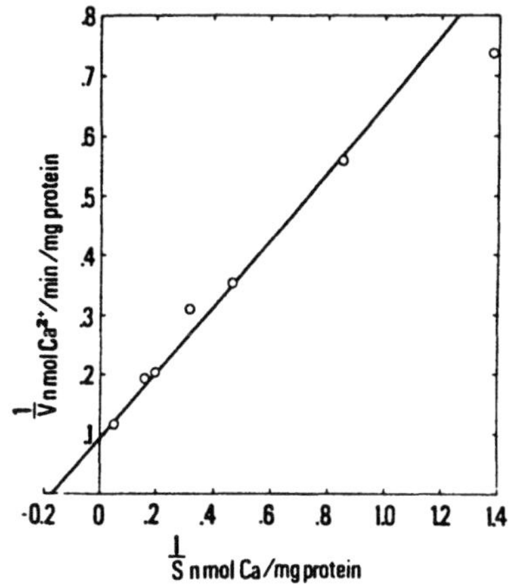

Fig. 13. The kinetics of release of Ca^{2+} from heart mitochondria at very low Ca loads. The rate of release of Ca^{2+} was measured using Arsenazo III and a dual-wavelength spectrophotometer. Release was initiated by addition of ruthenium red and NaCl to mitochondria respiring in the steady state. Prior to this the mitochondrial Ca content had been either supplemented by small additions of $CaCl_2$, or partially depleted by small additions of EGTA. Mitochondrial Ca content was determined by absorbance change on adding the ionophore A23187 or by atomic absorption spectrophotometry of mitochondrial extracts. These techniques gave identical results; this parameter is designated "S". For full experimental details, see the original paper. (*Hansford* and *Castro* 1982)

Coll et al. 1982) or a true asymmetry in the carrier or its environment on the two faces of the mitochondrial membrane.

Most importantly for this discussion, the $S_{0.5}$ value for heart mitochondrial Ca efflux is five times higher than the $K_{0.5}$ value for the activation by Ca^{2+} of 2-oxoglutarate dehydrogenase (Fig. 10) or pyruvate dehydrogenase interconversion (*Hansford* and *Castro*, unpublished). Both parameters were originally measured as total mitochondrial Ca, and thus the validity of the "activity coefficient" derived from the null-point data (*Hansford* and *Castro* 1982) does not affect this conclusion. Direct corroboration of this finding comes from the work of *Coll* et al. (1982) who determined $S_{0.5}$ for matrix Ca^{2+} to be 5.7 μM in studies of Ca^{2+} efflux and $K_{0.5}$ for Ca^{2+} to be 1.2 μM in studies of the activation of 2-oxoglutarate dehydrogenase in heart mitochondria. *Coll* et al. (1982) found an $S_{0.5}$ value of 9.7 μM for Ca^{2+} efflux from liver mitochondria. The conclusion from these results is that mitochondria containing a total Ca low enough to allow modulation of dehydrogenase activity (see Sect. 3.3) cannot serve accurately to buffer $[Ca^{2+}]c$. The reason for this is that accurate buffering demands that the Ca^{2+} efflux pathway be saturated: this is presented in Fig. 14.

Further experimentation on this point revealed that heart mitochondria exposed to a range of "loads" of total Ca exist in a steady state with the

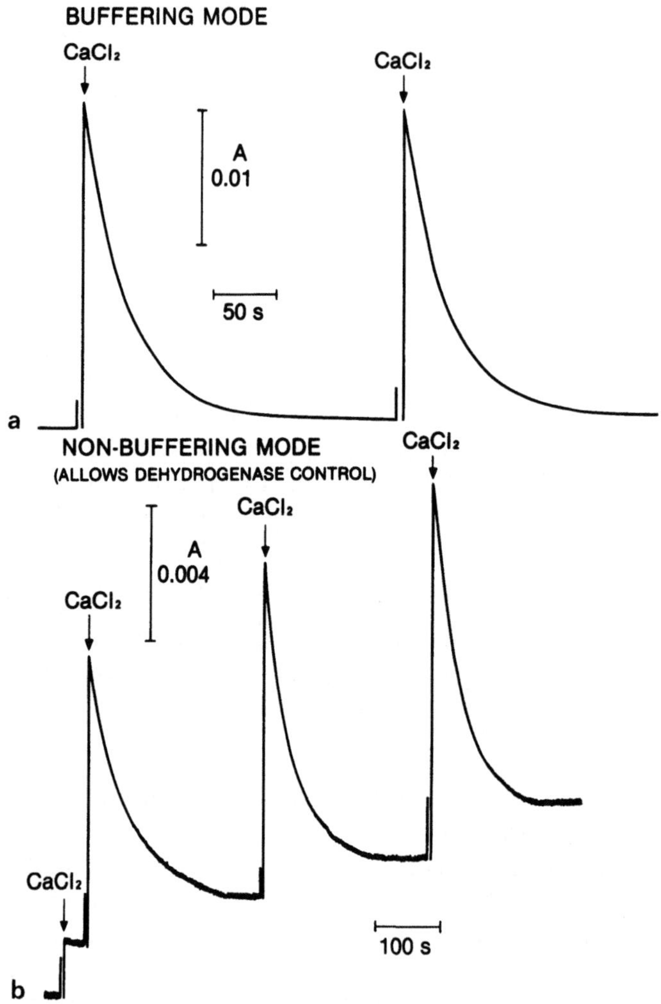

Fig. 14a,b. Mitochondrial Ca^{2+} transport: the buffering (**a**) and the nonbuffering (**b**) mode. Ca^{2+} uptake by heart mitochondria was monitored by using Arsenazo III and a dual-wavelength spectrophotometer. The medium comprised 0.12 *M* KCl, 20 m*M* K Hepes, 10 m*M* NaCl, 5 m*M* KPi, 4 m*M* L-glutamate, 5 m*M* L-malate, and 30 *μM* Arsenazo III. In **a**, 1 mg mitochondrial protein was used, and an addition of 10 nmol $CaCl_2$ had been made prior to the record presented. Where shown, 20 nmol additions of $CaCl_2$ were made. In **b**, 1 mg mitochondrial protein was used, and the mitochondria had been largely depleted of endogenous Ca by an initial addition of 20 nmol EGTA. Additions of $CaCl_2$ were each of 10 nmol: the first addition was largely bound by residual EGTA. Note that the sensitivity and time scale of the two recordings are different.

Rationale: In the *buffering mode* **a**, the mitochondria contained 12 nmol Ca/mg protein at the point of addition of the first pulse of $CaCl_2$. At this Ca content, the egress pathway is saturated (Fig. 13). Addition of 10 nmol $CaCl_2$ gives rise to net Ca accumulation; however, this does not raise the activity of the egress pathway, which remains at its former level. Therefore, uptake occurs until the activity of the uniporter also achieves its former value, and since this is controlled by $[Ca^{2+}]o$, this entails a return of $[Ca^{2+}]o$ to the value characteristic of the incubation before the pulse of same value of $[Ca^{2+}]o$, provided the mitochondrial total Ca content is 3 nmol/mg protein or more (Fig. 15). This is the region of perfect buffering emphasized by other investigators (*Bygrave* 1978; *Nicholls* 1978; *Nicholls* and *Crompton* 1980) and the $[Ca^{2+}]o$ value associated with this steady state is the so-called set-point (*Becker* et al. 1980; *Brand* and *De Selincourt* 1980). When the Ca content is less than 3 nmol/mg protein, considerable deviation from perfect buffering occurs (Fig. 15), a predicted consequence of the nonsaturation of the efflux pathway under these conditions. In this region, which the data in Sect. 4 and especially the results of electron-probe analysis suggest is the physiologically relevant region of Ca contents, at least for muscle mitochondria, the value of $[Ca^{2+}]o$ is neither "set" nor a "point." The term is catchy, but probably misleading, and should perhaps be abandoned.

The relationship between mitochondrial Ca content and $[Ca^{2+}]c$ in liver may be rather different from that in heart, as developed in an elegant, very recent paper by *Joseph* et al. (1983). The authors showed that in experiments with isolated liver mitochondria, the value of $[Ca^{2+}]o$ was quite constant at mitochondrial Ca contents of 4 nmol/mg protein or above, but declined when the total Ca content of the mitochondria was depleted below this value. This result is reminiscent of that presented for heart mitochondria above. However, this paper made the further advance that a model was developed, involving V_{max} and K_m parameters for influx and efflux pathways, which fully described the relationship between mitochondrial Ca content and $[Ca^{2+}]o$. The model allows the derivation of these kinetic parameters from the distribution of Ca^{2+} across the mitochondrial membrane in the steady state. Further, it identifies a change in the K_m of the Ca^{2+} uniporter as being the modification most capable of materially changing the steady-state distribution of Ca^{2+} across the mitochondrial membrane, as part of the mechanism of response to hormones. The difference from the situation in heart, described above, arises from studies which *Joseph* et al. (1983) carried out with isolated hepatocytes, in which they measured $[Ca^{2+}]c$ using the null-point method, and the distribution of Ca^{2+} between mitochondria and endoplasmic reticulum using release by uncoupling agent as a criterion of mitochondrial Ca and subsequent release by the ionophore A23187 as a criterion of reticulum Ca.

$CaCl_2$. In the ***nonbuffering mode* b**, the mitochondria contained less than 1 nmol Ca/mg protein prior to the first pulse of $CaCl_2$, which resulted in a cycle of uptake. At this Ca content the egress pathway is not saturated (Fig. 13). Consequently, the net accumulation generated by the addition of 10 nmol of $CaCl_2$ results in an increase in activity of the egress pathway, which is matched in the ensuing steady state by an increase in activity of the uptake (uniporter) pathway. This requires that the value of $[Ca^{2+}]o$ remain higher than it was prior to the pulse of $CaCl_2$

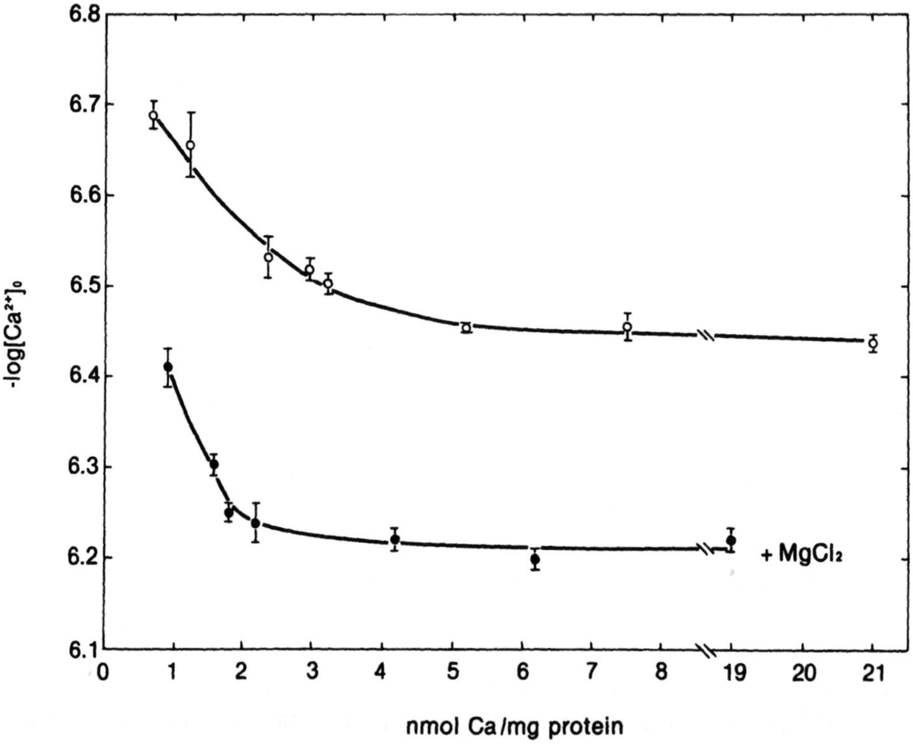

Fig. 15. The lack of buffering of $[Ca^{2+}]o$ by heart mitochondria at a low content of total Ca. $[Ca^{2+}]o$ was determined using a Ca^{2+}-specific electrode. Ca content was manipulated by small additions of $CaCl_2$ or EGTA, and was determined in mitochondrial extracts by atomic absorption spectrophotometry. Where indicated, $MgCl_2$ was present at 1 m*M*. Full experimental details are given in the original paper. (*Hansford* and *Castro* 1982)

The finding was that hepatocytes, as isolated, contain 5–6 nmol Ca/mg dry wt., under which circumstances the reticulum system is saturated with Ca and the mitochondria contain 16 nmol Ca/mg mitochondrial protein, a content high enough to saturate the mitochondrial efflux pathway and ensure coexistence with a $[Ca^{2+}]c$ of 0.3 μM. This presumably corresponds to the mitochondrial "set-point." These studies are coherent and convincing and led *Joseph* et al. (1983) to conclude that liver mitochondria operate in the buffering mode (Fig. 14) in the intact cell, whereas those in heart may operate in the mode of dehydrogenase control. This conclusion hangs entirely on the physiological relevance of cellular Ca contents of the magnitude seen in this study. As summarized above (Sect. 4.2), some studies of in situ Ca content of liver mitochondria have indicated values substantially lower (and some higher) than 5–6 nmol/mg dry wt. Conceivably, plasma membrane Ca^{2+} transport processes may be affected by the cell isolation procedure, allowing the artifactual loading of the cells with

Ca. In the face of these and other theoretical points concerning the primacy of the plasma membrane in cellular Ca^{2+} homeostasis (see below), it seems premature to conclude that the model of dehydrogenase control by Ca^{2+} does not apply to liver (*Joseph* et al. 1983).

The discussion so far has been couched in terms of the $K_{0.5}$ value for $[Ca^{2+}]$m for dehydrogenase activation and the $S_{0.5}$ value for $[Ca^{2+}]$m for Ca^{2+} efflux. The former is known only for heart and the latter only for heart and liver mitochondria (*Hansford* and *Castro* 1982; *Coll* et al. 1982). Although it is very likely that the $K_{0.5}$ for dehydrogenase activation by intramitochondrial Ca^{2+} will be similar in different tissues, as this is true for the $K_{0.5}$ Ca^{2+} of the isolated enzymes (*Denton* et al. 1980; *McCormack* and *Denton* 1980), it is quite possible that the kinetics of the Ca^{2+}-efflux pathway may differ. Thus, the effect of Na^+ (*Crompton* et al. 1978; *Goldstone* et al. 1983) and membrane potential (*Zoccarato* and *Nicholls* 1982) on Ca^{2+} efflux is different in liver and heart. The relationship between these two parameters, i.e., $K_{0.5}$ for dehydrogenase activation and $S_{0.5}$ for Ca^{2+} efflux, will determine whether there is any overlap between the mode of dehydrogenase control and that of $[Ca^{2+}]$c buffering in any given tissue.

Recently, *Zoccarato* and *Nicholls* (1982) have emphasized the role of inorganic phosphate in binding intramitochondrial Ca^{2+} and ensuring that the activity of Ca^{2+} efflux is constant at varying Ca contents, though not saturated. They infer that the $[Ca^{2+}]$m in liver is in the low μM range in the presence of physiologically appropriate concentrations of inorganic phosphate, this inference being based on the activity of the efflux pathway in the presence of phosphate and of acetate. Clearly this situation would allow the perfect buffering of $[Ca^{2+}]$c, the mechanism being the sequestering of a proportion of the total CaPi in an insoluble form. However, it is not at all clear that it would also allow modulation of dehydrogenase activity, as suggested by *Zoccarato* and *Nicholls* (1982), as the $[Ca^{2+}]$m would be effectively clamped in the presence of CaPi in excess of its solubility product and a constant concentration of Pi. A sine qua non for the control of enzyme activity by Ca^{2+} is an ability of $[Ca^{2+}]$m to be changed. In addition, such a preeminent role of Pi did not emerge in studies in which $[Ca^{2+}]$m was actually measured in heart mitochondria (*Hansford* and *Castro* 1982, and unpublished), in which this parameter was very insensitive to Pi or acetate. This may have been a consequence of the low but physiologically appropriate total mitochondrial Ca contents investigated (0.5–5 nmol/mg mitochondrial protein).

6.2 A Simple Model

It seems to the author that a simple model consistent with most experimental findings is that mitochondrial Ca^{2+}-transport mechanisms exist to translate changes in $[Ca^{2+}]c$ into changes in $[Ca^{2+}]m$, and that the advantage this confers to the animal is that increased energy demands, often signalled by increased $[Ca^{2+}]c$, are met by increased fluxes in catabolic metabolism, with improved homeostasis of tissue ATP/ADP × Pi ratios. The former part of this model is presented in a simple, qualitative, and intuitive form in Fig. 16; the latter part will be examined in Sect. 8.

Let us consider three instances in which $[Ca^{2+}]c$ may change, and examine possible effects on $[Ca^{2+}]m$. In Fig. 16a, a transient increase in plasma membrane Ca^{2+} permeability gives an increase in $[Ca^{2+}]c$. An example might be the depolarization of a presynaptic nerve ending (*Blaustein* 1975). In the long term, an equal amount of Ca^{2+} has to be pumped out of the cell to allow the maintenance of a steady state (*Schweitzer* and *Blaustein* 1980). In the short term, there is a degree of buffering both by the mitochondria and, perhaps more importantly, by an endoplasmic reticulum fraction (*Schweitzer* and *Blaustein* 1980; but, compare *Nicholls* and *Scott* 1980; *Åkerman* and *Nicholls* 1983). The uptake of Ca^{2+} by the mitochondria will be very dependent on the value of $[Ca^{2+}]c$ generated on depolarization, owing to the sigmoidal V versus S dependence and high V_{max} of the Ca^{2+} uniporter in mitochondria from nervous tissue (*Brinley* et al. 1978). Values of $[Ca^{2+}]c$ in the activated state are not known for presynaptic nerve endings, but in vitro studies with synaptosomes have indicated a substantial mitochondrial Ca content on depolarization (*Åkerman* and *Nicholls* 1981a). On the present model, dehydrogenase activity will increase with depolarization, assuming that the $[Ca^{2+}]m$ in vivo is not saturating in the resting state. Some evidence in favor of this assumption derives from pyruvate dehydrogenase activation on depolarization (see Sect. 7.2). This being so, it is unlikely that the neuronal mitochondrial efflux system is saturated with Ca^{2+}; unpublished work by *Hansford* and *Castro* indicates 50% of V_{max} activity at a mitochondrial Ca content of 50 nmol/mg protein, in rat brain mitochondria incubated in a buffered isotonic medium containing 1 m*M* $MgCl_2$, 1 m*M* ATP and 5 m*M* Pi. Thus, the activity of the efflux pathway will rise, generating a new, higher, set-point. The mitochondria have no primary role in setting the value for $[Ca^{2+}]c$, however, as, to return the tissue to the resting steady state, the plasma membrane must be capable of decreasing the $[Ca^{2+}]c$ below the new, elevated mitochondrial set-point and thus of depleting the mitochondria of Ca^{2+}. This depletion would lead to the inactivation of the dehydrogenase systems under discussion. Provocatively, there would be an expected lag in dehydrogenase inactivation, owing to the relative slowness

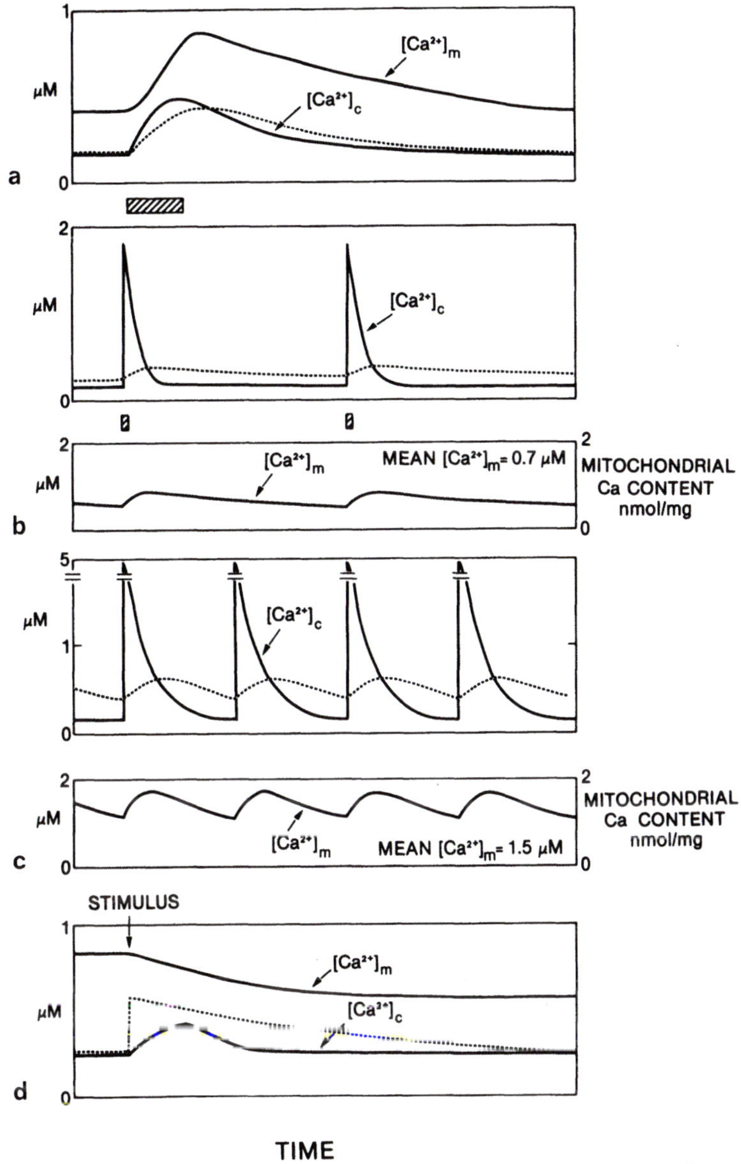

Fig. 16a–d. A simple model of relations between mitochondrial Ca content, $[Ca^{2+}]m$, and $[Ca^{2+}]c$. **a** The response to a transient influx of Ca^{2+} across the plasma membrane (duration indicated by *hatched bar*). **b** The response to periodic brief episodes of Ca^{2+} entry across the plasma membrane. The *lower panel* presents mitochondrial Ca content and $[Ca^{2+}]m$. **c** The response to periodic plasma membrane Ca^{2+} entry of a higher frequency than that shown in **b**. **d** The response to a hormone-induced decrease in the ratio of activities of uniporter/antiporter. In each panel, anticipated values of the "set-point" are presented by a *dotted line*. Absolute values of all parameters presented are arbitrary assumptions. For discussion, please see the text

of the mitochondrial efflux pathways (relative to influx), just as there may be a lag in inactivation of the cytosolic Ca^{2+}-calmodulin–dependent enzymes, owing to the hysteretic behavior of calmodulin-Ca^{2+} binding (*Rasmussen* and *Waisman* 1983).

Figures 16b and c present variants of Fig. 16a in which repetitive stimulation and plasma membrane Ca^{2+} influx occurs, as in heart. Fig. 16c differs from 16b in that it assumes a higher frequency of plasma membrane depolarization and Ca^{2+} influx. As discussed in Section 5.1, the mitochondrial Ca content and $[Ca^{2+}]$m may be expected to be heavily weighted by the relatively very high values of $[Ca^{2+}]$c attained during contraction (*Marban* et al. 1980) and the potentially very high activity of the Ca^{2+} uniporter (*Scarpa* and *Graziotti* 1973). Efflux is less rapid but can proceed for a longer time, as $[Ca^{2+}]$c has a value of $< 10^{-7}$ *M* for the major part of the heart cycle (*Allen* and *Blinks* 1978). The degree of damping of changes in $[Ca^{2+}]$m relative to $[Ca^{2+}]$c in Figs. 16b and c is quite hypothetical and unknown. However, it is clear that the mean value of $[Ca^{2+}]$m will be higher in the situation shown in Fig. 16c than that shown in 16b.

Figure 16d represents an entirely different situation, with a putative primary effect of a hormone upon a mitochondrial transport process. This possibility has been widely raised in connection with the mode of action of α_1-agonists on liver cells (see Sect. 7.3). The recent modeling study of *Joseph* et al. (1983) has identified the K_m for Ca^{2+} of the uniporter as being the parameter most sensitive to change, with the system being relatively insensitive to changes in the kinetic parameters of the antiporter. Consistent with this is a recent experimental demonstration by *Zoccarato* and *Nicholls* (1982) showing a rather modest change in $[Ca^{2+}]$o of a suspension of liver mitochondria when efflux rates were varied 16-fold. If the uniporter were to be inhibited in response to a hormone second messenger, the changes envisaged in Fig. 16d might occur. In a qualitative way, the set-point might be expected to change somewhat as shown, bringing into question the degree to which it is set. The suggestion that the $[Ca^{2+}]$c would return to the value prior to the intervention (Fig. 16d), demonstrating a preeminent role of plasma membrane transport, is consistent with what is seen when a major portion of the cytosolic Ca^{2+} of lymphocytes is bound to the indicator Quin 2: the $[Ca^{2+}]$c regains its value prior to the perturbation (*Tsien* et al. 1982). In truth, Fig. 16d may be only a paper exercise, as there is some evidence that the Ca^{2+}-sensitive dehydrogenases are activated by α_1-agonists (see Sect. 7.3.1), consistent with the portion of the literature which shows an increased, rather than a decreased, mitochondrial Ca content in response to these agents (see Sect. 7.3.2 for references). This may be a consequence of an increased $[Ca^{2+}]$c (see below), with possibly a contribution from an enhanced activity of the Ca uniport, as has been shown by *Kessar* and *Crompton* (1981) to be a result of the exposure of heart to the α-agonist methoxamine. In the event of a decreased K_m for the uniporter in response to a hormonal second messenger, one might expect the converse of the behavior shown in Fig. 16c, namely a maintained increase in mitochondrial Ca and a transient decrease in $[Ca^{2+}]c$.

7 The Physiology of Dehydrogenase Regulation by Ca^{2+}

This section discusses the evidence that control of mitochondrial dehydrogenases by Ca^{2+} ions actually functions in intact tissues. The best evidence comes from excitable tissues, and these will be discussed first.

7.1 The Role of Ca^{2+} in the Response of Cardiac PDH_A Content to Increased Work Load and β-Adrenergic Stimulation

7.1.1 Background

Increased cardiac work performance is associated with an increase in the amount of pyruvate dehydrogenase present as PDH_A (*Illingworth* and *Mullings* 1976; *Hiltunen* and *Hassinen* 1976; *Reinauer* and *Muller-Ruchholtz* 1976; *Pearce* et al. 1980; *McCormack* and *Denton* 1981a; *Bünger* et al. 1982). Associated with the increased work performance is a decrease in the tissue ATP/ADP ratio (*Illingworth* et al. 1975; *Reinauer* and *Muller-Ruchholtz* 1976) and a decrease in the mitochondrial $NADH/NAD^+$ ratio. The attribution of the latter change to the mitochondrial compartment can be made either by using the glutamate dehydrogenase equilibrium as an indicator (*Hiltunen* and *Hassinen* 1976) or by minimizing the contribution of cytosolic NADH to the fluorescence of heart tissue by the use of pyruvate as substrate (*Illingworth* et al. 1975). These changes in the ratios of effectors of pyruvate dehydrogenase interconversion would alone be expected to generate an increased PDH_A content (see reviews by *Hansford* 1980; *Reed* 1981; *Wieland* 1983). However, the possibility that another factor may be involved emerges from a study of the time course of the changes reported by *Illingworth* et al. (1975) and *Illingworth* and *Mullings* (1976). Thus, PDH_A content continues to increase over the period from 30 s to 1 min after the imposition of a work load, whereas the ATP/ADP and $NADH/NAD^+$ ratios undergo no further change after 30 s. A plausible mechanism for the continued increase in PDH_A content beyond the 30 s point is that mitochondrial Ca content, $[Ca^{2+}]m$, and pyruvate dehydrogenase phosphatase activity are increased, thus generating a raised steady state PDH_A content. This is credible if there is a rather substantial lag in the transmission of changes in $[Ca^{2+}]c$ into changes in $[Ca^{2+}]m$.

Clearly, with the proven relationship between $[Ca^{2+}]c$, force development, and actomyosin ATPase in muscle (*Katz* 1970), an increase in $[Ca^{2+}]c$ is expected to go hand in hand with a decrease in the ratios of mitochondrial [ATP]/[ADP] and [NADH]/$[NAD^+]$. Until very recently, distinguishing the direct effects of Ca^{2+} on pyruvate dehydrogenase interconversion from indirect effects due to these nucleotide ratios proved impossible. This emerged, for instance, in attempts to define the mechanism whereby β-adrenergic stimulation leads to an elevated $[PDH_A]$ in heart (*Hiraoka* et al. 1980; *McCormack* and *Denton* 1981a; *Bünger* et al. 1982). *Hiraoka* et al. (1980) clearly demonstrated that changes in $[Ca^{2+}]c$ are involved in the response to epinephrine and isoproterenol by showing that the increases in PDH_A content are blocked by the Ca^{2+} channel–blocker verapamil (see *Fleckenstein* 1977) and are dependent on the perfusate Ca^{2+} concentration in the presence of the bivalent metal ionophore A23187. Such findings are in line with the general understanding that β-agonists raise $[Ca^{2+}]c$ in heart muscle (*Williamson* 1975), but do not allow a distinction to be made between direct effects of changes in $[Ca^{2+}]m$ upon pyruvate dehydrogenase interconversion, and indirect effects mediated by changes in the mitochondrial [ATP]/[ADP] and [NADH]/$[NAD^+]$ ratios (see above).

7.1.2 Recent Definitive Experiments

Such a distinction between direct and indirect effects upon pyruvate dehydrogenase interconversion has in fact been made, however, in very recent experiments which took advantage of the ability of ruthenium red to inhibit mitochondrial Ca^{2+} uptake (*Moore* 1971). It was found that perfusion of rat hearts with medium containing ruthenium red completely abolished an increase in PDH_A content caused either by high perfusate Ca^{2+} concentrations or by the β-agonist isoproterenol (*McCormack* and *England* 1983). These results are shown in Fig. 17a,b. The work is greatly strengthened by the additional measurement of contractile force and the tissue content of phosphorylase *a*, as these measurements indirectly report on changes in $[Ca^{2+}]c$ (*Katz* 1970; *Brostrom* et al. 1971; *Ozawa* 1972). It is seen (Fig. 17b) that there are concentrations of ruthenium red which totally abolish the response of PDH_A content to β-agonists or to elevated perfusate Ca^{2+} concentrations, while causing no change in phosphorylase *a* content and only minimal changes in contractile force. Further, and very importantly, ruthenium red does not at all diminish the response of contractile force to the positive inotropic intervention. From the study of *McCormack* and *England* (1983) it is clear that perfusion with epinephrine,

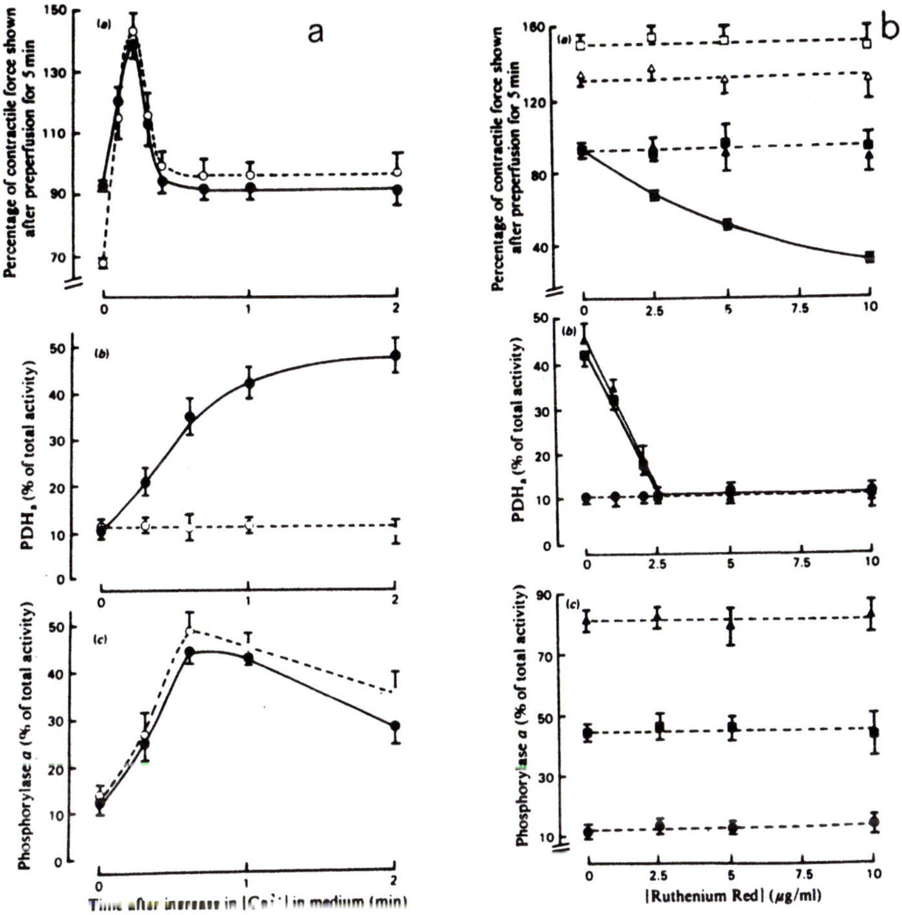

Fig. 17a,b. The effect of ruthenum red on the increase in PDH_A content seen when perfused heart is exposed to elevated perfusion fluid Ca^{2+} concentration or to isoproterenol. **a** presents the response of ***PDH_A content, force production,*** and ***phosphorylase a content*** to an elevation in perfusion fluid Ca^{2+} concentration: ●, control; ○, plus ruthenium red. Ruthenium red uniquely abolishes the response of PDH_A content. **b** demonstrates that the increase in PDH_A content owing to elevated Ca^{2+} concentration or to isoproterenol is more sensitive to inhibition by ruthenium red than is the increase in force production and in phosphorylase *a* content, i.e., the parameters responsive to $[Ca^{2+}]c$. (*McCormack* and *England* 1983)

isoproterenol, or elevated Ca^{2+} concentrations gives an increased $[Ca^{2+}]c$; this affects pyruvate dehydrogenase interconversion via a ruthenium red–sensitive process, presumably mitochondrial Ca^{2+} uptake, and the response of pyruvate dehydrogenase is predominantly mediated via changes in $[Ca^{2+}]m$, and not through indirect mechanisms involving [ATP]/[ADP] and $[NADH]/[NAD^+]$. The conclusion follows from the fact that there is no correspondence between contractile force and PDH_A content in the presence of ruthenium red, although increased contractile force would be expected to be accompanied by decreased mitochondrial ratios of [ATP]/[ADP] and [NADH]/[NAD$^+$]. Finally, one more relevant finding in this important paper is that ruthenium red does not affect the PDH_A content of the resting heart (Fig. 17b). Clearly, mitochondrial Ca^{2+} cycling will occur under these conditions, i.e., even at very low mitochondrial Ca contents (*Nicholls* 1978; *Hansford* and *Castro* 1982), so the only obvious way of rationalizing this finding is to say that the activation of the pyruvate dehydrogenase phosphatase by Ca^{2+} is negligible in the resting state.

Confirmation of the primacy of changes in [Ca^+]m in mediating the response of heart PDH_A content to epinephrine comes from some very recent studies by *McCormack* and *Denton* (1984), in which mitochondria were extracted from hearts which had been perfused either under control conditions, in the presence of epinephrine, or at elevated perfusate Ca^{2+} concentrations. Quite remarkably, examination of the PDH_A content of these mitochondria revealed increases due to the positive inotropic interventions which were stable during isolation. Figure 18 presents results for PDH_A content during the isolation procedure and during subsequent incubation of the mitochondria. The secret of the success in preserving these Ca^{2+}- or hormone-induced changes in PDH_A content was the exclusion of Na^+ from the medium: addition of Na^+ caused a rapid relaxation of the raised PDH_A content to control values, an effect which was blocked by the Ca^{2+}-channel-blocker diltiazem. Diltiazem has been shown to inhibit

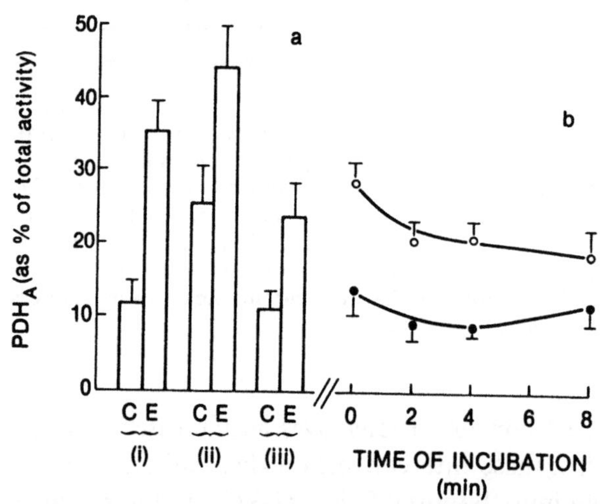

Fig. 18a,b. The increase in PDH_A content seen when hearts are perfused with epinephrine is persistent during subsequent mitochondrial isolation. In **a**, the effect of epinephrine on PDH_A content is presented for (*i*) the freeze-clamped heart, (*ii*) the freshly prepared homogenate, and (*iii*) the subsequent mitochondrial fraction. Epinephrine-treated tissue is denoted by *E* and control tissue by *C*. **b** illustrates the stability of the hormone-induced change in mitochondria incubated at 30° for the times indicated. ○, Epinephrine; ●, control. (Redrawn from *McCormack* and *Denton* 1984)

$Ca^{2+}/2Na^+$ antiport in heart mitochondria (*Vághy* et al. 1982). Thus there is a strong inference that the preservation of an elevated PDH_A content is associated with the preservation of a raised mitochondrial Ca content (and $[Ca^{2+}]m$) during isolation. Indeed, an analogy was demonstrated with the maintenance of ^{45}Ca content of isolated heart mitochondria, loaded in vitro to the point at which pyruvate dehydrogenase was largely activated. Here, too ^{45}Ca content was remarkably preserved during a reisolation of the mitochondria, but rapidly discharged in the presence of Na^+; the additional presence of diltiazem prevented this discharge.

This work by *McCormack* and *Denton* (1984) is the most direct evidence we have that inotropic stimulation of heart by epinephrine and β-agonists gives rise to increased PDH_A content via an increase in mitochondrial Ca content – and the associated parameter $[Ca^{2+}]m$. The conclusion is further strengthened by their finding that 2-oxoglutarate dehydrogenase activity, measured at nonsaturating 2-oxoglutarate concentrations, is also elevated in mitochondria derived from epinephrine-treated hearts. The response of this enzyme to mitochondrial Ca and $[Ca^{2+}]m$ is described in Section 3.3 (and see *Hansford* and *Castro* 1981). Thus, these two papers (*McCormack* and *England* 1983; *McCormack* and *Denton* 1984) provide excellent evidence for a primary control of pyruvate dehydrogenase interconversion (and 2-oxoglutarate dehydrogenase activity) by $[Ca^{2+}]m$ in the response of the heart to β-adrenergic stimulation.

7.2 The Role of Ca^{2+} in the Response of PDH_A Content to the Stimulation of Nervous Tissue

It has recently become clear that long-lasting changes in the content of PDH_A may form part of the mechanism whereby synaptic transmission is potentiated in response to repetitive stimulation (*Browning* et al. 1979, 1981a) or to a training experience (*Morgan* and *Routtenberg* 1981). Provocatively, the change in phosphorylation of pyruvate dehydrogenase seen on repetitive stimulation of hippocampal slices is abolished if the bathing medium is Ca-free (*Browning* et al. 1979). At the same time, preganglionic stimulation is known to increase the Ca content of superior cervical ganglia, as measured by either ^{45}Ca (*Blaustein* 1971) or by electron-probe X-ray microanalysis (*Parducz* and *Joó* 1976). The question arises as to whether an enhanced entry of Ca^{2+} into the mitochondria is responsible for generating the increased PDH_A content seen on repetitive stimulation. At the moment this cannot be answered with certainty, as the entry of Ca^{2+} into the presynaptic terminal may also trigger ATPase activity and thus give rise to decreased mitochondrial ratios of ATP/ADP and $NADH/NAD^+$ – in analogy with the consequences of a raised work load imposed upon the heart, as discussed above (Sect. 7.1). Perhaps an elevated neuronal mitochondrial Ca content would survive the isolation of mitochondria from stimulated hippocampal slices, provided that the isolation medium is Na^+ free, as the Ca egress pathway of nervous tissue mitochondria is known to be Na^+ dependent (*Crompton* et al. 1978). This, then would be the counterpart of the experiment carried out on heart tissue by *McCormack* and *Denton* (1984).

What is clear is that an increased PDH_A content in response to electrical stimulation may have important consequences for synaptic transmission. The argument has been made that increased pyruvate dehydrogenase activity will increase the activity of mitochondrial Ca^{2+} uptake and thus alter the buffering of cytosolic free Ca^{2+} concentrations in nervous tissue (*Browning* et al. 1981b). While model experiments did indeed show that Ca^{2+} uptake by brain mitochondria is potentiated by the presence of dichloroacetate (which has the effect of raising PDH_A content via inhibition of the kinase) when the only respiratory substrate available is a non-saturating concentration of pyruvate (*Browning* et al. 1981b), it seems unlikely that Ca^{2+} uptake is limited by substrate oxidation in vivo. The reason for this assertion is the finding by *Nicholls* (1978) that substantial decreases in mitochondrial membrane potential, $\Delta\psi$, are necessary before Ca^{2+} uniport activity is compromised. Thus, the ability of the mitochondria to maintain the phosphorylation potential of cytosolic adenine nucleotides at a level compatible with life would seem to guarantee undiminished rates of Ca^{2+} uniport. In fact, the experimental system used by *Browning* et al. (1981b) differs from the physiological situation in the absence of oxidizable substrates other than pyruvate, e.g., glutamate.

However, it seems to the reviewer that these demonstrations of changes in PDH_A content in response to electrical stimulation (*Browning* et al. 1979, 1981a) or to training (*Morgan* and *Routtenberg* 1981) are very exciting in the context of the potential for neurotransmitter synthesis. Thus, a proven consequence of an increased PDH_A content is an increased mitochondrial acetyl-CoA/CoASH ratio, especially when the activity of the respiratory chain is restrained by physiologically appropriate ATP/ADP ratios (*Hansford* and *Cohen* 1978). As acetyl-CoA is the acetyl donor for the synthesis of acetylcholine, as indicated by the requirement for flux through pyruvate dehydrogenase (*Gibson* et al. 1975), this increase in acetyl-CoA/CoASH ratio may well be associated with increased rates of acetylcholine synthesis. Thus one can speculate, with some plausibility, that long-term potentiation of synapses involves increased $[Ca^{2+}]m$, increased PDH_A content, increased steady-state acetyl-CoA/CoASH ratios, and consequent increased neurotransmitter synthesis. Equally, the impact of senescence upon brain function, which has been shown to involve a less active synthesis and release of acetylcholine (*Gibson* and *Peterson* 1981), could derive from a decrease in $[Ca^{2+}]m$.

Fig. 19. The effect of treating hepatocytes with vasopressin on PDH_A content. PDH_A content is presented as a function of vasopressin concentration. For experimental details, see the original paper. (*Denton* et al. 1981)

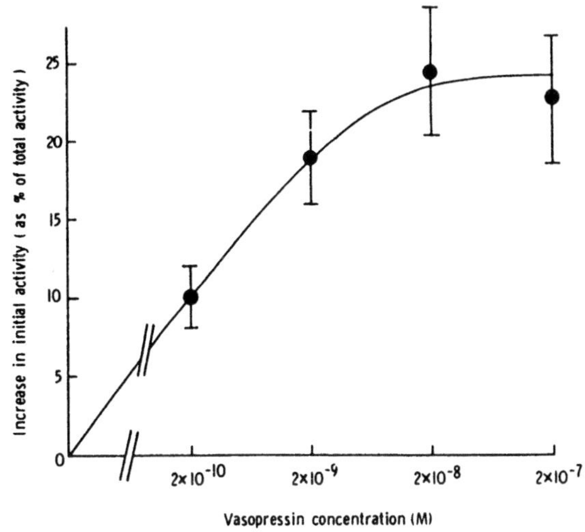

7.3 The Response of Liver Ca^{2+}-Sensitive Dehydrogenases to α_1-Adrenergic Stimulation, Vasopressin, and Angiotensin II

7.3.1 Evidence for Dehydrogenase Activation by Ca^{2+}

This section covers the considerable, albeit indirect, evidence in favor of the notion that epinephrine, α_1-agonists, and vasopressin activate liver pyruvate and 2-oxoglutarate dehydrogenases through an elevation of $[Ca^{2+}]m$.

Exposure of the perfused liver (*Hems* et al. 1978; *Sies* et al. 1983) or isolated hepatocytes (*Denton* et al. 1981) to vasopressin results in an increase in the content of PDH_A, as shown in Fig. 19. Exposure of hepatocytes to the α-agonist phenylephrine has a similar effect (*Assimacopoulos-Jeannet* et al. 1983). These effects are not likely to be mediated by changes in the ratios of the known effectors of interconversion, as the ATP/ADP ratio is unchanged (*Hems* et al. 1978; *Assimacopoulos-Jeannet* et al. 1983) or increased (*Siess* et al. 1978) in response to vasopressin. Nor is a change in pyruvate concentration part of the mechanism, as this decreases in response to the hormone (*Hems* et al. 1978). In view of the requirement for extracellular Ca^{2+} for an effect of vasopressin to be shown on PDH_A content, a reasonable interpretation of these results is that the hormone acts to increase $[Ca^{2+}]m$ and thus potentiate pyruvate dehydrogenase phosphatase activity (*Hems* et al. 1978; *Denton* et al. 1981).

Similar conclusions concerning $[Ca^{2+}]m$ emerge from studies of gluconeogenesis and flux through the tricarboxylate cycle. Thus, vasopressin stimulates the oxidation of oleate by hepatocytes in the presence but not in the absence of extracellular Ca^{2+} (*Sugden* et al. 1980a,b). The effect is localized to the tricarboxylate cycle by the demonstration that a simultaneous inhibition of ketogenesis occurs, and to the level of the 2-oxoglutarate dehydrogenase reaction by the demonstration that the increased flux is associated with a decrease in the steady-state concentration of 2-oxoglutarate (*Sugden* et al. 1980a). Further, in fractionation studies of hepatocytes *Siess* et al. (1978) showed that vasopressin, as well as epinephrine and the α-agonist phenylephrine, gave rise to decreased steady-state concentrations of 2-oxoglutarate in the mitochondrial compartment. As they also showed an increase in the mitochondrial $NADH/NAD^+$ ratio (*Siess* et al. 1978), a strong case can be made for the occurrence of an allosteric activation of 2-oxoglutarate dehydrogenase in response to vasopressin and phenylephrine; this is evidenced by simultaneous increase in flux and decrease in the concentrations of the reactants 2-oxoglutarate and NAD^+ (see *Rolleston* 1972 for an evaluation of metabolite "crossover" data). Similar conclusions emerge from the report of *Taylor* et al. (1983), showing that enhanced fatty acid oxidation, flux through 2-oxoglutarate dehydrogenase, and β-hydroxybutyrate/acetoacetate ratios accompany the treatment of perfused liver from starved rats with the α-agonist phenylephrine. A requirement for the demonstration of these effects is the presence of a pool of cellular Ca, which can be depleted by successive stimulation by α-agonists at low extracellular Ca^{2+} concentrations (*Taylor* et al. 1983). In view of this, it is perhaps reasonable to think that the allosteric activator of 2-oxoglutarate dehydrogenase recognized above is in fact the Ca^{2+} ion.

An important part of the evidence in favor of dehydrogenase activation during stimulation of liver by vasopressin and by α_1-agonists is the increase in mitochondrial NADH content, which has been observed by means of fluorescence studies (*Sugano* et al. 1980; *Balaban* and *Blum* 1982; *Charest* et al. 1983) and by measurement of the β-hydroxybutyrate/acetoacetate ratio (*Siess* et al. 1978). The increase in NADH is transient, though an eventual steady-state $NADH/NAD^+$ ratio higher than that prior to the addition of the hormone may be achieved (*Sugano* et al. 1980). Accompanying this increased reduction of nicotinamide nucleotide is an increased rate of O_2 uptake, which is maintained as long as the hormone or agonist is present. This has been shown to accompany the response of hepatocytes to vasopressin (*Balaban* and *Blum* 1982) and phenylephrine (*Charest* et al. 1983) and of perfused liver to phenylephrine, vasopressin, and angiotensin II (*Reinhardt* et al. 1982a,b). Although the increased rates of O_2 uptake could theoretically be driven either by an increase in mitochondrial $NADH/NAD^+$ ratio or by a decrease in adenine nucleotide phosphorylation potential (*van der Meer* et al. 1978; *Wilson* et al. 1977), the experimental results clearly show that the former is involved. This follows from the demonstrated increase in mitochondrial $NADH/NAD^+$ ratio (*Siess* et al 1978; *Sugano* et al. 1980) and the lack of change of phosphorylation potential (*Hems* et al. 1978; *Assimacopoulos-Jeannet* et al. 1983). This identifies the activation of dehydrogenases catalyzing rate-limiting non-equilibrium reactions as being part of the mechanism of action of these hormones on liver tissue.

The tentative identification of the activator as being Ca^{2+} depends upon evidence that the response to vasopressin or α_1-agonists is lost after repeated stimulation in low-Ca^{2+} media, but not in normal-Ca^{2+} media (*Taylor* et al. 1983) and that a correlate of stimulation by these compounds is a rise in $[Ca^{2+}]c$, as demonstrated either by increases in the content of phosphorylase *a* (*Murphy* et al. 1980; *Chrisman* et al. 1982) or directly, using either the fluorescent Ca^{2+} indicator Quin 2 (*Charest* et al. 1983) or the "null-point" method (*Murphy* et al. 1980). Studies with isolated mitochondria in turn indicate that such an increase in $[Ca^{2+}]c$ within the range 0.2–0.4 μM (*Murphy* et al. 1980; *Charest* et al. 1983) will generate an increased steady-state mitochondrial Ca content (*Nicholls* 1978), and that this is associated with increases in $[Ca^{2+}]m$ (*Coll* et al. 1982). Accompanying such an increase is an increased PDH_A content and 2-oxoglutarate dehydrogenase activity, albeit in tissues other than liver (see Sect. 3.3).

Finally, the lag observed in recent, elegant experimentation by *Reinhart* et al. (1982a) and by *Charest* et al. (1983) between α-agonist-induced Ca^{2+} mobilization, as judged either by increases in $[Ca^{2+}]c$ or by enhanced efflux from liver tissue, and increases in O_2 uptake (*Reinhart* et al. 1982a) or NADH fluorescence (*Charest* et al. 1983) is entirely consistent with a mechanism whereby an increased $[Ca^{2+}]c$ gives rise, secondarily, to an increased $[Ca^{2+}]m$ and to dehydrogenase activation. According to this interpretation, the lag of 2.5 s (*Reinhart* et al. 1982a) or 4.5 s (*Charest* et al. 1983) reflects the response of mitochondrial Ca^{2+} cycling to the new value of $[Ca^{2+}]c$.

7.3.2 Results of the Direct Measurement of Mitochondrial Ca Content in Response to α_1-Agonists, Vasopressin

Although Sect. 7.3.1 provides quite compelling, albeit indirect, evidence that vasopressin and α_1-agonists increase $[Ca^{2+}]m$ in liver, it has to be admitted that there is also a body of experimental work leading to the opposite conclusion. Much of the earlier work can be criticized on the grounds that perfusion media contained subphysiological concentrations of Ca^{2+} (*Chen* et al. 1978; *Blackmore* et al. 1979b; *Babcock* et al. 1979; *Althaus-Salzmann* et al. 1980), thereby depleting cellular Ca pools. In addition, earlier tissue fractionation studies leading to the identification of changes in Ca content of mitochondrial fractions are open to criticism on the grounds that isolation times were lengthy, or inhibitors of both uniport and antiport pathways were not present, thus possibly allowing redistribution of Ca^{2+} to take place (*Blackmore* et al. 1979b; *Poggioli* et al. 1980). However, there have been recent methodological advances leading to more convincing findings on cellular Ca^{2+} distribution. Thus, *Reinhart* et al. (1982b) demonstrated an α-agonist-induced loss of Ca from a liver mitochondrial fraction isolated in less than 6 min in the presence of both ruthenium red and dibucaine (Nupercaine) to prevent Ca^{2+} redistribution. Although the absolute Ca content of this fraction was dependent on the perfusate Ca^{2+} concentration, being approximately 2.7 nmol/mg protein at 10 μM and 6.8 nmol/mg protein at 1.3 mM perfusate Ca^{2+}, the magnitude of the α-agonist-induced loss was similar in each case, at around 0.8 nmol Ca/mg protein.

Equally convincing was the demonstration by *Murphy* et al. (1980) that a mitochondrial fraction derived from norepinephrine-treated hepatocytes by a technique involving disruption by digitonin and shearing forces followed by rapid centrifugation contained less Ca than a similar fraction from control cells. Agreement between mitochondrial Ca content thus determined and that determined on the basis of releasability on FCCP treatment of the intact cells lends credence to this report.

Clearly, a decreased mitochondrial Ca content, as shown by *Murphy* et al. (1980) and *Reinhart* et al. (1982b), is not consistent with the well-described increase in $[Ca^{2+}]c$ which accompanies stimulation by α-agonists and vasopressin (*Murphy* et al. 1980; *Charest* et al. 1983: see *Exton* 1981; *Williamson* et al. 1981 for reviews) unless some change in the kinetics of the mitochondrial Ca^{2+}-uniport or -antiport pathways takes place. Otherwise, an increased $[Ca^{2+}]c$ or $[Ca^{2+}]o$ would lead to an increased mitochondrial Ca content, whether the preexisting mitochondrial Ca was in the range allowing buffering, as described by *Nicholls* (1978), or in the lower range allowing less than complete buffering as identified by *Hansford* and *Castro* (1982).

Experiments with mitochondria isolated from livers treated with phenylephrine have revealed somewhat modest changes in the kinetics of the efflux pathway, with a twofold decrease in K_m for intramitochondrial Ca^{2+} and no change in V_{max} (*Coll* et al. 1982). The consequence of the decrease in K_m may be offset by the finding that the so-called activity coefficient of intramitochondrial Ca is also decreased as a result of this intervention. In any event, if the model of primary control of $[Ca^{2+}]c$ by the mitochondria has validity, the kinetic parameter most responsive to hormonal modulation is the K_m of the Ca^{2+} uniporter (*Joseph* et al. 1983). So far, there has been no convincing evidence of α-adrenergic effects on the Ca^{2+} uniporter of liver mitochondria. The apparent activation of uniport activity reported by *Taylor* et al. (1980) and by *Reinhart* et al. (1980) needs reassessment, as measurements were made at concentrations of Ca^{2+}

at which transport is probably limited by proton extrusion by the respiratory chain (see also *Åkerman* and *Nicholls* 1983). Certainly, there has been no report of a stable inhibition of uniport activity, as would be required to explain the diminished mitochondrial Ca content of α-agonist-treated liver found by *Murphy* et al. (1980) and *Reinhart* et al. (1982b). *Coll* et al. (1982) have speculated that an α-agonist-induced increase in $[Mg^{2+}]c$ would give the prerequisite inhibition of uniport activity in situ: there is, however, no evidence for such an increase (*Corkey* and *Hale* 1982).

Stable changes in either Ca^{2+} transport pathway remain an attractive possible mechanism of hormone action, with an activation of the uniporter occurring in heart mitochondria as a consequence of α-adrenergic stimulation (*Kessar* and *Crompton* 1981), and an activation of the (inferred) antiporter of liver mitochondria occurring as a consequence of β-adrenergic stimulation (*Goldstone* and *Crompton* 1982; *Goldstone* et al. 1983). However, to date, such changes provide no explanation of the apparent release of mitochondrial Ca by α-agonists (*Murphy* et al. 1980; *Reinhart* et al. 1982b).

Thus, our picture of the effect of α-agonists on cellular Ca^{2+} distribution and, of central relevance to this article, on the parameter $[Ca^{2+}]m$ is still far from complete. The case made by *Murphy* et al. (1980), *Reinhart* et al. (1982b), *Blackmore* et al. (1982), and *Morgan* et al. (1983) that these agents interact with the mitochondria, presumably though a second messenger, alter the relative activities of uniport and antiport, and thus cause mitochondrial Ca^{2+} release is clearly consistent with some recent experimental evidence. Nevertheless, there are some loose ends. *Blackmore* et al. (1982) have shown that oligomycin acts in the same sense as mitochondrial uncoupling agents in mobilizing cell Ca, i.e., in increasing $[Ca^{2+}]c$, and that this is closely related to the depletion of cellular ATP. Similarly, *Morgan* et al. (1983) have shown that oligomycin is as effective as uncoupling agents or respiratory inhibitors in preventing a large increase in hepatocyte Ca content otherwise seen on stimulation with both α-agonists and glucagon. In this case, the mitochondrial accumulation of Ca occurring in parallel with cellular accumulation was also abolished by oligomycin. The authors interpret these results as being in accord with a primary effect of these hormones on mitochondrial, and perhaps also reticular, Ca^{2+} transport. It seems to the reviewer that these experiments instead establish the role of the mitochondrion as providing ATP for some other membrane Ca^{2+} transport system, quite possibly in the endoplasmic reticulum. Oligomycin would not be expected to inhibit mitochondrial Ca^{2+} uptake per se, as it does not diminish the driving force, $\Delta\psi$, and indeed may even increase this quantity (*Johnson* and *Hansford* 1977). Oligomycin does, however, prevent mitochondrial synthesis of ATP.

Thus, it is the bias of the reviewer that a reticular pool of Ca^{2+} is perhaps a more likely target for α-adrenergic release than is a mitochondrial pool. A contribution of reticulum to net release was acknowledged in the work of *Reinhart* et al. (1982b) on the grounds of inadequacy of the mitochondrial pool to account for total cellular release. Further, a loss of Ca from a microsomal fraction was shown to occur. Very recently, the quantitative aspects of hormone-stimulated Ca^{2+} release from both mitochondria and reticulum have been addressed by *Joseph* and *Williamson* (1983) in a study of hepatocytes which used releasability by uncoupling agents as a criterion of mitochondrial location. These authors found that there is no release whatsoever from the mitochondrial compartment in response to phenylephrine when the Ca content of the mitochondria is less than 1.8 nmol/mg mitochondrial protein. Under these conditions, Ca^{2+} is mobilized exclusively from the reticulum. However, at higher Ca contents release from both organelles occurs, with the mitochondrion potentially playing a larger role owing to the saturation characteristics and relatively small capacity of the reticular system in liver (*Joseph* et al. 1983; *Joseph* and *Williamson* 1983). The release of mitochondrial Ca at the higher loads, in the face of the known increase in $[Ca^{2+}]c$ (*Murphy* et al. 1980; *Charest* et al. 1983), requires a decrease in the activity ratio of uniporter to antiporter due to the hormonal intervention. The lack of release of mitochondrial Ca at the lower values of mitochondrial Ca content then suggests that this change in the activity ratio is offset by the rise in $[Ca^{2+}]c$, resulting in no change in mitochondrial Ca. The increase in $[Ca^{2+}]c$ is attributable to release from the reticulum. The lack of change of mitochondrial Ca content at the lower values of cellular Ca shown in this study (*Joseph* and *Williamson* 1983) should not be taken to rule out an increase in $[Ca^{2+}]m$ in response to α_1-stimulation at these lower values of cellular Ca. This is because the techniques used, of Ca^{2+} release by uncoupling agents and ionophores, do not really have the sensitivity to describe changes at the lower values of mitochondrial Ca content (below 1.8 nmol/mg mitochondrial protein), which are relevant to the model of dehydrogenase activation. What this very recent contribution does highlight is the need to establish exactly what the liver tissue Ca content and mitochondrial Ca content are in situ, in the living animal. Only then will it be possible to evaluate whether the mitochondrion loses or gains Ca in response to these hormones.

Finally, it seems very hard to explain the fascinating recent results of *Morgan* et al. (1983) on interactions between α_1- and β-adrenergic activation on the basis of a primary site of action at the level of mitochondrial Ca^{2+} transport. These authors found that, whereas α_1- or β-activation singly caused a release of Ca^{2+} from hepatocytes and perfused liver, these agonists in concert promoted a very large uptake, including uptake into a mitochondrial fraction. These results are anomalous if discussed in terms of a primary effect on mitochondrial Ca^{2+} cycling: in these terms, α_1-agonists would have to both decrease the activity ratio of uniporter to antiporter to allow the Ca^{2+} mobilization characteristic of α-agonists, and increase this ratio in the presence of cAMP to potentiate the net mitochondrial uptake seen under those conditions. Clearly, puzzles remain.

7.4 The Response of Ca^{2+}-Sensitive Dehydrogenases to Stimulation by Other Hormones

The β-adrenergic stimulation of heart results in an increased content of PDH_A, and this may reasonably be linked to Ca^{2+} fluxes into the cell, and thence to the mitochondria. This has been discussed in Section 7.1.

By contrast, there appears to be no evidence that β-adrenergic stimulation increases the content of PDH_A in liver, and most work has indicated that glucagon, which also serves to increase tissue cAMP content, is without effect (*Patzelt* et al. 1973; *Claus* and *Pilkis* 1977; *Denton* et al. 1981). Given the premise of this article this is somewhat surprising, as glucagon has been shown to lower the Ca content of mitochondria in liver tissue in situ (*Friedmann* and *Rasmussen* 1970; *Chen* et al. 1978; *Blackmore* et al. 1979b; *Whiting* and *Barritt* 1982; *Baddams* et al. 1983). However, as discussed above, the pyruvate dehydrogenase interconversion system is complex, and effects mediated by changes in $[Ca^{2+}]m$ could be masked by changes in the other effector ratios. Quite significantly, in their recent work *Baddams* et al. (1983) have shown that a prior perfusion of liver with glucagon decreases the Ca^{2+} that can subsequently be released from intracellular stores by epinephrine. A decreased mitochondrial Ca pool would be predicted to lead to lower activities of Ca^{2+}-sensitive dehydrogenase systems according to the bias of this article, and indeed there is some evidence for decreased activity of 2-oxoglutarate dehydrogenase in response to glucagon from the studies of tricarboxylate cycle flux in hepatocytes carried out by *Sugden* et al. (1980b). The finding by *Morgan* et al. (1983) that mitochondrial Ca content is actually enhanced in liver by the simultaneous presence of glucagon and α_1-agonists or vasopressin guarantees that the metabolic implications of these two hormones acting singly and in concert will be quite complex. Notably, against the background of reports showing no effect of glucagon on liver PDH_A content, there is one recent report showing a marked increase in the presence of both glucagon and phenylephrine (*Assimacopoulos-Jeannet* et al. 1983). This, then, is in accord with the Ca^{2+}-distribution studies of *Morgan* et al. (1983), on the basis of mitochondrial Ca being a prime effector of pyruvate dehydrogenase interconversion.

Until very recently, it seemed plausible that the increase in PDH_A content brought about by exposure of epididymal adipose tissue to insulin (*Jungas* 1970; *Denton* et al. 1971; *Weiss* et al. 1971; *Denton* and *Hughes* 1978) might be mediated by a rise in $[Ca^{2+}]m$. This suggestion arose because of the persistence of the effect during mitochondrial isolation and because of the implication of a change in phosphatase, as opposed to kinase, activity (*Severson* et al. 1976; *Hughes* and *Denton* 1976). However, recent experiments, analogous to those discussed in Sect. 7.1.2 for heart mitochondira, have shown that the inhibitor ruthenium red prevents the increase in PDH_A content of fat pads incubated at a high medium Ca^{2+} concentration (5 m*M*), but has no effect on the increase caused by exposure to insulin (*Marshall* et al. 1984). Further, the increased PDH_A content due to insulin treatment persisted when mitochondria were isolated from the fat pads and incubated in media containing Na^+, conditions which poteniate Na^+/Ca^{2+} antiport and which were shown to lead to the discharge of accumulated ^{45}Ca. Finally, when the PDH_A content of mitochondria from insulin-treated pads is compared with that of control mitochondria as a function of $[Ca^{2+}]o$, it is found that PDH_A values for the insulin-treated tissue are higher across the whole range of $[Ca^{2+}]o$, i.e., $10^{-6}-10^{-8}$ *M*. Were the mechanism of the insulin effect an enhanced mitochondrial Ca content, this would be expected to be generated in the mitochondria from control tissues simply by raising the $[Ca^{2+}]o$, and this is not seen. Curiously, however, the $K_{0.5}$ value for activation by extramitochondrial Ca^{2+} does seem to be lower in the mitochondria from insulin-treated tissues (e.g., 419 ± 20 n*M* for the control and 229 ± 47 n*M* for the insulin-treated, in the presence of 10 m*M* NaCl and 1 m*M* $MgCl_2$). This is an anomaly, as such an altered sensitivity to extramitochondrial $[Ca^{2+}]$ might be expected to give rise to different mitochondrial Ca contents in insulin-treated and control tissues, on the assumption that $[Ca^{2+}]c$ is the same. Nevertheless, the work by *Marshall* et al. (1984) is in most respects a convincing proof that the increased PDH_A content of insulin-treated fat pads is not associated with an elevated value of $[Ca^{2+}]m$. This conclusion is buttressed by their finding that there is no activation of 2-oxoglutarate dehydrogenase by insulin which is stable to mitochondrial isolation: such a consequence of β-adrenergic stimulation of heart has been shown to survive mitochondrial isolation (*McCormack* and *Denton* 1984).

The lack of involvement of changes in $[Ca^{2+}]m$ in mediating the effect of insulin on PDH_A content means that the provocative studies of *Jarett* and *Seals* (1979), *Seals* and *Czech* (1980), and *Kiechle* et al. (1981) on a putative second messenger of insulin action fall beyond the scope of this review.

8 Some Thoughts on the Biological Advantages of Dehydrogenase Level Control

Rates of electron transfer down the mitochondrial respiratory chain are driven by the disequilibrium between ΔEh and Δp, according to the chemiosmotic theory of energy coupling (*Mitchell* 1966, 1979), where ΔEh is the difference in observed reduction potentials across a span of the chain and Δp is the proton-motive force, or electrochemical potential difference for protons across the mitochondrial inner membrane. Rates of phosphoryl group transfer are similarly driven by the disequilibrium between Δp and ΔGATP, where the latter is the free energy of ATP synthesis under prevailing cellular conditions (see *Hansford* 1980 for a review). The increased reduction of NAD^+ by 2-oxoglutarate (*Hansford* and *Castro* 1981), occurring as a consequence of dehydrogenase activation upon the availability of Ca^{2+} to heart mitochondria, serves to increase the thermodynamic driving force for oxidative phosphorylation. For a contractile or other excitable tissue, metabolic control should probably be thought of in terms of the provision of both ADP and Ca^{2+} to the mitochondria, rather than in terms of the availability of ADP alone as envisaged in the classic theory of *Chance* and *Williams* (1956). Inspection of results obtained with heart mitochondria oxidizing nonsaturating concentrations of 2-oxoglutarate indicates that the Eh of the NAD^+/NADH and cytochrome c Fe^{3+}/Fe^{2+} couples changes very little in the transition from a resting state [minus ADP, minus Ca^{2+}] to an active state [plus ADP, plus Ca^{2+}] (*Hansford* and *Castro* 1981). Equally, there is very recent work by *Kauppinen* et al. (1984) showing that the Eh of the mitochondrial NAD^+/NADH couple is essentially unchanged in the intact heart on the transition from a beating state to KCl-induced arrest. By contrast, the classic result obtained in mitochondrial experiments with a substrate showing no dehydrogenase-level activation is a pronounced oxidation of the NAD^+/NADH couple upon the provision of ADP (*Chance* and *Williams* 1956). Thus, the activation of the 2-oxoglutarate dehydrogenase by Ca^{2+} and by ADP (*McCormack* and *Denton* 1979) offers the biological advantage of an undiminished thermodynamic driving force despite the increase in flux through oxidative phosphorylation. This allows high rates of oxidative phosphorylation without the necessity of lowering the adenine nucleotide phosphorylation potential to levels incompatible with the ATP-requiring processes of the cytosol. In insect flight muscle, a tissue uniquely adapted to high rates of energy transduction, activation at the dehydrogenase level is even clearer and results in an overall reduction of NAD and of cytochrome c on the transition from rest to flight (*Keilin* 1925; *Hansford* 1980). In this case, the important modulators are Ca^{2+}, activating the glycerol 3-phosphate dehydrogenase (see Sect. 2.5) and ADP, activating the NAD-isocitrate dehydrogenase (Sect. 2.2).

Although steady-state levels of reduction of respiratory chain components have not been measured as a function of Ca^{2+} activation of NAD-isocitrate dehydrogenase, it seems very likely that a picture similar to that seen with 2-oxoglutarate dehydrogenase would emerge. This is based upon the increased reduction of NAD and cytochrome c seen when the insect NAD-isocitrate dehydrogenase is activated by ADP, which lowers the apparent K_m for isocitrate, in analogy to the effect of Ca^{2+} on the enzyme from higher animals (*McCormack* and *Denton* 1981b).

By contrast, *Hansford* (1981) showed that activation of pyruvate dehydrogenase in heart leads to no significant increase in reduction of NAD in state-4 (*Chance* and *Williams* 1956) respiration. Thus, the argument concerning an increased thermodynamic driving force seems not to apply to this substrate oxidation. The reason for the difference may be a stringent end-product inhibition by NADH, acting at the level of the enzymes of interconversion (see *Hansford* and *Cohen* 1978). However, this is not to diminish the importance of increases in PDH_A content consequent on Ca^{2+} availability. In respiratory states intermediate between 3 and 4 (see *Chance* and *Williams* 1956, for terminology), which probably are most appropriate to physiology, an increased PDH_A content may well allow an increased steady-state $NADH/NAD^+$ ratio. Equally important, the elevated [acetyl-CoA]/[CoASH] ratio which is a consequence of a high PDH_A content and flux-limitation by the respiratory chain (*Hansford* and *Cohen* 1978) may have important consequences for neurotransmitter synthesis (see Sect. 7.2) and for intermediary metabolism in liver. Thus, the increased PDH_A content characteristic of liver activated by α_1-agonists and vasopressin (see Sect. 7.3 for references) would be predicted to lead to an increased mitochondrial [acetyl]/[CoASH] ratio – this apparently has not been measured – with consequences for ketogenesis and gluconeogenesis.

Finally, in the discussion of the biological significance of dehydrogenase activation by Ca^{2+} ions, it has to be emphasized that higher animals have intramitochondrial dehydrogenase systems activated by Ca^{2+} and lower animals do not (*McCormack* and *Denton* 1981b). Equally, higher animals have active mitochondrial Ca^{2+} transport pathways and lower organisms do not. Of the organisms surveyed by *McCormack* and *Denton* (1981b), only locust flight muscle, blowfly flight muscle, potato, and *E. coli* dehydrogenases showed little or no activation by Ca^{2+}. The uptake of Ca^{2+} by both insect flight muscle (*Carafoli* et al. 1971) and plant (*Carafoli* and *Lehninger* 1971; *Moore* and *Bonner* 1977) mitochondria is extremely sluggish and, in the case of the insect, is seen only at concentrations of Ca^{2+} implausibly high for the cytosol (*Carafoli* et al. 1971; *Bygrave* et al. 1975). The one mitochondrial dehydrogenase in insect flight muscle which is activated by Ca^{2+} (*Hansford* and *Chappell* 1967) faces outward on the mitochondrial membrane (*Klingenberg* and *Buchholz* 1970) and therefore does not depend on transmembrane ion transport. Thus, the conclusion that mitochondrial Ca^{2+} transport systems serve to modulate the parameter [Ca^{2+}]m and control catabolic metabolism is hard to avoid.

9 Summary

The three nonequilibrium reactions involved in the complete oxidation of pyruvate and acetyl groups (the pyruvate dehydrogenase system, NAD-isocitrate dehydrogenase, and 2-oxoglutarate dehydrogenase) are located within the mitochondrial matrix space and are activated by Ca^{2+}, responding in intact, respiring heart and skeletal muscle mitochondria to changes in [Ca^{2+}]o plausible for the cytosol in vivo.

Modulation of dehydrogenase activity in muscle mitochondria requires that total mitochondrial Ca content be 2 nmol/mg protein or less, corresponding to approximately 2 μM free matrix Ca^{2+} ([Ca^{2+}]m) or less. There is evidence, especially from electron-probe analysis, that this is true of muscle mitochondria in situ.

At this value of [Ca^{2+}]m the antiporter catalyzing Ca^{2+} efflux from these mitochondria is not saturated, precluding precise buffering of the free Ca^{2+} concentration of the mitochondrial surroundings ([Ca^{2+}]c or [Ca^{2+}]o). Thus, the cycling of Ca^{2+} across the inner mitochondrial membrane in muscle cells does not serve to accurately buffer [Ca^{2+}]c but serves instead to relay changes in [Ca^{2+}]c into changes in [Ca^{2+}]m.

This regulates dehydrogenase activity, allowing increased energy demands signalled through increases in [Ca^{2+}]c to be met by increased rates of oxidative phosphorylation, without the necessity for a large decrease in adenine nucleotide phosphorylation potential. Such an adjustment of PDH_A content and 2-oxoglutarate dehydrogenase activity to increased work load clearly occurs in heart.

β-Adrenergic stimulation of heart activates the pyruvate dehydrogenase system and 2-oxoglutarate dehydrogenase through an increase in [Ca^{2+}]c translated into an increase in [Ca^{2+}]m.

Essentially similar arguments apply to mitochondria from nervous tissue, which also exhibit increased dehydrogenase activity in response to increased [Ca^{2+}]o in the range $10^{-6}-10^{-7}$ M. Physiologically, depolarization of presynaptic nerve endings is associated with an increased cellular Ca content and an increased PDH_A content, with possible implications for acetylcholine synthesis.

Isolated liver mitochondria do not show changes in dehydrogenase activity with exogenous substrates when the $[Ca^{2+}]o$ is varied from 10^{-6} to 10^{-7} *M*, possibly a function of limitation by membrane permeability to oxidizable substrate rather than by dehydrogenase activity. The Ca content found in many, but not all, liver mitochondrial preparations may be too high to allow modulation of dehydrogenase activity; this is not yet clear.

α_1-Adrenergic stimulation of liver and treatment with vasopressin increase PDH_A content and $[Ca^{2+}]c$. The relationship between the two parameters is not entirely clear, with a body of evidence favoring a release of mitochondrial Ca^{2+} due to the hormones. If this is so, the mediator of the increased PDH_A content has still to be described. An alternative explanation is that the Ca^{2+} is mobilized from the endoplasmic reticulum, giving an increase in $[Ca^{2+}]c$ and allowing, secondarily, an increase in $[Ca^{2+}]m$. This would then provide a mechanism for the increased PDH_A content.

Dehydrogenase activity could also be controlled by changes in the relative activities of mitochondrial Ca^{2+} uniport and antiport pathways, in the absence of changes in $[Ca^{2+}]c$. Such changes have been reported with α- and β-adrenergic stimulation, but so far the data are too incomplete to explain the action of either group of agonists.

The thesis that the mitochondrial Ca^{2+} transport cycle exists to modulate intramitochondrial metabolism is strengthened by the finding of a correspondence between the possession of high-affinity Ca^{2+} transport systems and the possession of Ca^{2+}-sensitive intramitochondrial dehydrogenases, in a study of different phyla.

Acknowledgments. Reproduction of previously published figures is with the permission of the copyright holders. I would like to thank the authors cited in the figure legends for permission to reproduce their work, and Drs. Denton and McCormack, and Kauppinen, Hiltunen, and Hassinen, especially, for their willingness to share their work with me prior to publication.

This article covers work published up to and including 1983.

References

Akerboom TPM, Bookelman H, Zuurendonk PF, van der Meer R, Tager JM (1978) Intramitochondrial and extramitochondrial concentrations of adenine nucleotides and inorganic phosphate in isolated hepatocytes from fasted rats. Eur J Biochem 84:413–420

Åkerman KEO, Nicholls DG (1981a) Intrasynaptosomal compartmentation of calcium during depolarization-induced calcium uptake across the plasma membrane. Biochim Biophys Acta 645:41–48

Åkerman KEO, Nicholls DG (1981b) Calcium transport by intact synaptosomes. Influence of ionophore A23187 on plasma-membrane potential, plasma-membrane calcium transport, mitochondrial membrane potential, respiration, cytosolic free-calcium concentration and noradrenaline release. Eur J Biochem 115:67–73

Åkerman KEO, Nicholls DG (1983) Physiological and bioenergetic aspects of mitochondrial calcium transport. Rev Physiol Biochem Pharmacol 95:149–201

Allen DG, Blinks JR (1978) Calcium transients in aequorin-injected frog cardiac muscle. Nature 273:509–513

Althaus-Salzmann M, Carafoli E, Jakob A (1980) Ca^{2+}, K^+ redistribution and α-adrenergic activation of glycogenolysis in perfused rat livers. Eur J Biochem 106: 241–248

Aogaichi T, Evans J, Gabriel J, Plaut GWF (1980) The effects of calcium and lanthanum ions on the activity of bovine heart nicotinamide adenine dinucleotide-specific isocitrate dehydrogenase. Arch Biochem Biophys 204:350–360

Assimacopoulos-Jeannet F, McCormack JG, Jeanrenaud B (1983) Effect of phenylephrine on pyruvate dehydrogenase activity in rat hepatocytes and its interaction with insulin and glucagon. FEBS Lett 159:83–88

Babcock DF, Chen J-LJ, Yip BP, Lardy HA (1979) Evidence for mitochondrial localization of the hormone-responsive pool of Ca^{2+} in isolated hepatocytes. J Biol Chem 254:8117–8120

Baddams HM, Chang LBF, Barritt GJ (1983) Evidence that glucagon acts on the liver to decrease mitochondrial calcium stores. Biochem J 210:73–77

Baker PF (1972) Transport and metabolism of calcium ions in nerve. Prog Biophys Mol Biol 24:177–223

Balaban RS, Blum JJ (1982) Hormone-induced changes in NADH fluorescence and O_2 consumption of rat hepatocytes. Am J Physiol 242:C172–C177

Barritt GJ, Lamont SV (1982) Kinetic properties of exchangeable calcium in guinea-pig heart mitochondria measured at low concentrations of free calcium and in the presence of Mg^{2+}, ATP^{4-} and inorganic phosphate. Cell Calcium 3:215–225

Barritt GJ, Spiel PF (1981) Effects of glucagon on ^{45}Ca outflow exchange in the isolated perfused rat heart. Biochem Pharmacol 30:1407–1414

Becker GL, Fiskum G, Lehninger AL (1980) Regulation of free Ca^{2+} by liver mitochondria and endoplasmic reticulum. J Biol Chem 255:9009–9012

Blackmore PF, Dehaye J-P, Strickland WG, Exton JH (1979a) α-Adrenergic mobilization of hepatic mitochondrial calcium. FEBS Lett 100:117–120

Blackmore PF, Dehaye J-P, Exton JH (1979b) Studies on α-adrenergic activation of hepatic glucose output. The role of mitochondrial calcium release in α-adrenergic activation of phosphorylase in perfused rat liver. J Biol Chem 254:6945–6950

Blackmore PF, Hughes BP, Shuman EA, Exton JH (1982) α-Adrenergic activation of phosphorylase in liver cells involves mobilization of intracellular calcium without influx of extracellular calcium. J Biol Chem 257:190–197

Blaustein MP (1971) Preganglionic stimulation increases calcium uptake by sympathetic ganglia. Science 172:391–393

Blaustein MP (1975) Effects of potassium, veratridine and scorpion venom on calcium accumulation and transmitter release by nerve terminals in vitro. J Physiol 247:617-655

Blaustein MP, Ector AC (1976) Carrier-mediated sodium-dependent and calcium-dependent calcium efflux from pinched-off presynaptic nerve terminals (synaptosomes) in vitro. Biochim Biophys Acta 419:295–308

Blinks JR, Wier WG, Hess P, Prendergast FG (1982) Measurement of Ca^{2+} concentrations in living cells. Prog Biophys Mol Biol 40:1–114

Borle AB (1981) Control, modulation, and regulation of cell calcium. Rev Physiol Biochem Pharmacol 90:13–153

Borle AB, Snowdowne KW (1982) Measurement of intracellular free calcium in monkey kidney cells with aequorin. Science 217:252–254

Borst P (1962) The pathway of glutamate oxidation by mitochondria isolated from different tissues. Biochim Biophys Acta 57:256–269

Bragadin M, Pozzan T, Azzone GF (1979) Kinetics of Ca^{2+} carrier in rat liver mitochondria. Biochemistry 18:5972–5978

Brand MD, De Selincourt C (1980) Effects of glucagon and Na^+ on the control of extra-mitochondrial free Ca^{2+} concentration by mitochondria from liver and heart. Biochem Biophys Res Commun 92:1377–1382

Brinley Jr FJ, Tiffert T, Scarpa A (1978) Mitochondria and other calcium buffers of squid axon studied in situ. J Gen Physiol 72:101–127

Brostrom CO, Hunkeler FL, Krebs EG (1971) The regulation of skeletal muscle phosphorylase kinase by Ca^{2+}. J Biol Chem 246:1961–1967

Browning M, Dunwiddie T, Bennett W, Gispen W, Lynch G (1979) Synaptic phosphoproteins: specific changes after repetitive stimulation of the hippocampal slice. Science 203:60–62

Browning M, Bennett WF, Kelly P, Lynch G (1981a) Evidence that the 40,000 Mr phosphoprotein influenced by high frequency synaptic stimulation is the alpha subunit of pyruvate dehydrogenase. Brain Res 218:255–266

Browning M, Baudry M, Bennett WF, Lynch G (1981b) Phosphorylation-mediated changes in pyruvate dehydrogenase activity influence pyruvate-supported calcium accumulation by brain mitochondria. J Neurochem 36:1932–1940

Bünger R, Permanetter B, Sommer O, Yaffe S (1982) Adaptive changes of pyruvate oxidation in perfused heart during adrenergic stimulation. Am J Physiol 242: H30–H36

Bygrave FL (1977) Mitochondrial calcium transport. Current Topics in Bioenergetics. Ed. DR Sanodi. Academic Press 6:259–318

Bygrave FL (1978) Mitochondria and the control of intracellular calcium. Biological Reviews of the Cambridge Philosophical Society 53:43–79

Bygrave FL, Daday AA, Doy FA (1975) Evidence for a calcium-ion-transport system in mitochondria isolated from flight muscle of the developing sheep blowfly *Lycilia cuprina*. Biochem J 146:601–608

Carafoli E (1967) In vivo effect of uncoupling agents on the incorporation of calcium and strontium into mitochondria and other subcellular fractions of rat liver. J Gen Physiol 50:1849–1864

Carafoli E, Lehninger AL (1971) A survey of the interaction of calcium ions with mitochondria from different tissues and species. Biochem J 122:681–690

Carafoli E, Hansford RG, Sacktor B, Lehninger AL (1971) Interaction of Ca^{2+} with blowfly flight muscle mitochondria. J Biol Chem 246:964–972

Chance B, Williams GR (1956) The respiratory chain and oxidative phosphorylation. Adv Enzymol 17:65–134

Chaplain RA (1967) Insect actomyosin ATP-ase – a regulatory protein. Arch Biochem Biophys 121:154–168

Chappell JB (1964) The effects of 2,4-dinitrophenol on mitochondrial oxidations. Biochem J 90:237–248

Charest R, Blackmore PF, Berthon B, Exton JH (1983) Changes in free cytosolic Ca^{2+} in hepatocytes following α_1-adrenergic stimulation. Studies on Quin-2-loaded hepatocytes. J Biol Chem 258:8769–8773

Chen J-LJ, Babcock DF, Hardy HA (1978) Norepinephrine, vasopressin, glucagon, and A23187 induce efflux of calcium from an exchangeable pool in isolated rat hepatocytes. Proc Natl Acad Sci USA 75:2234–2238

Chrisman TD, Jordan JE, Exton JH (1982) Purification of rat liver phosphorylase kinase. J Biol Chem 257:10798–10804

Claret-Berthon B, Claret M, Mazet JL (1977) Fluxes and distribution of calcium in rat liver cells: kinetic analysis and identification of pools. J Physiol 272:529–552

Claus TH, Pilkis SJ (1977) Effect of dichloroacetate and glucagon on the incorporation of labelled substrates into glucose and on pyruvate dehydrogenase in hepatocytes from fed and starved rats. Arch Biochem Biophys 182:52–63

Coll KE, Joseph SK, Corkey BE, Williamson JR (1982) Determination of the matrix free Ca^{2+} concentration and kinetics of Ca^{2+} efflux in liver and heart mitochondria. J Biol Chem 257:8696–8704

Cooper RH, Randle PJ, Denton RM (1974) Regulation of heart muscle pyruvate dehydrogenase kinase. Biochem J 143:625–641

Corkey BE, Hale DE (1982) Magnesium distribution in cytosol and mitochondria of hepatocytes isolated from normal and diabetic rats. Fed Proc 41:1457

Crompton M, Heid I (1978) The cycling of calcium, sodium and protons across the inner membrane of cardiac mitochondria. Eur J Biochem 91:599–608

Crompton M, Sigel E, Salzmann M, Carafoli E (1976a) A kinetic study of the energy-linked influx of Ca^{2+} into heart mitochondria. Eur J Biochem 69:429–434

Crompton M, Capano M, Carafoli E (1976b) The sodium-induced efflux of calcium from heart mitochondria. A possible mechanism for the regulation of mitochondrial calcium. Eur J Biochem 69:453–462

Crompton M, Kunzi M, Carafoli E (1977) The calcium-induced and sodium-induced effluxes of calcium from heart mitochondria. Eur J Biochem 79:549–558

Crompton M, Moser R, Lüdi H, Carafoli E (1978) The interrelations between the transport of sodium and calcium in mitochondria of various mammalian tissues. Eur J Biochem 82:25–31

Crompton M, Heid I, Baschera C, Carafoli E (1979) The resolution of calcium fluxes in heart and liver mitochondria using the lanthanide series. FEBS Lett 104:352–354

Davis EJ, Lumeng L (1975) Relationships between the phosphorylation potentials generated by liver mitochondria and respiratory state under conditions of adenosine diphosphate control. J Biol Chem 250:2275–2282

deHaan EJ, Tager JM, Slater EC (1967) Factors affecting the pathway of glutamate oxidation in rat-liver mitochondria. Biochim Biophys Acta 131:1–13

Denton RM, Hughes WA (1978) Pyruvate dehydrogenase and the hormonal regulation of fat synthesis in mammalian tissues (Minireview). Int J Biochem 9:545–552

Denton RM, McCormack JG (1980) On the role of the calcium transport cycle in heart and other mammalian mitochondria. FEBS Lett 119:1–8

Denton RM, McCormack JG (1981) Calcium ions, hormones and mitochondrial metabolism. Clin Sci 61:135–140

Denton RM, Coore HG, Martin BR, Randle PJ (1971) Insulin activates pyruvate dehydrogenase in rat epididymal adipose tissue. Nature 231:115–116

Denton RM, Randle PJ, Martin BR (1972) Stimulation by Ca^{2+} of pyruvate dehydrogenase phosphate phosphatase. Biochem J 128:161–163

Denton RM, Randle PJ, Bridges BJ, Cooper RH, Kerbey AL, Pask HT, Severson DL, Stansbie D, Whitehouse S (1975) Regulation of mammalian pyruvate dehydrogenase. Mol Cell Biochem 9:27–53

Denton RM, Richards DA, Chin JG (1978) Calcium ions and the regulation of NAD^+-linked isocitrate dehydrogenase from the mitochondria of rat heart and other tissues. Biochem J 176:899–906

Denton RM, McCormack JG, Edgell NJ (1980) Role of calcium ions in the regulation of intramitochondrial metabolism. Effects of Na^+, Mg^{2+} and ruthenium red on the Ca^{2+}-stimulated oxidation of oxoglutarate and on pyruvate dehydrogenase activity in intact rat heart mitochondria. Biochem J 190:107–117

Denton RM, McCormack JG, Oviasu OA (1981) Short-term regulation of pyruvate dehydrogenase activity in the liver. In: Hue L, van de Werve G (eds) Short-term regulation of liver metabolism. Elsevier, Amsterdam, pp 159–174

DiPolo R, Requena J, Brinley JR FJ, Mullins LJ, Scarpa A, Tiffert T (1976) Ionized calcium concentration in squid axons. J Gen Physiol 67:433–467

Exton JH (1981) Molecular mechanisms involved in α-adrenergic responses. Mol Cell Endocrinol 23:233–264

Ezawa I, Ogata E (1977) Ca^{2+} requirement in ATP-induced activation of uncoupled oxidation of succinate in isolated rat-liver mitochondria. Eur J Biochem 77:427–435

Ezawa I, Ogata E (1979) Ca^{2+}-induced activation of succinate dehydrogenase and the regulation of mitochondrial oxidative reactions J Biochem 85:65–74

Fisher AB, Scarpa A, LaNoue KF, Bassett D, Williamson JR (1973) Respiration of rat lung mitochondria and the influence of Ca^{2+} on substrate utilization. Biochemistry 12:1438–1445

Fiskum G, Lehninger AL (1982) Mitochondrial regulation of intracellular calcium. In: Calcium and cell function. Vol 2, Ed. WY Cheung. New York, Academic, pp 39–80

Fleckenstein A (1977) Specific pharmacology of calcium in myocardium, cardiac pacemakers, and vascular smooth muscle. Annu Rev Pharmacol Toxicol 17:149–166

Foden S, Randle PJ (1978) Calcium metabolism in rat hepatocytes. Biochem J 170: 615–625

Foldes M, Barritt GJ (1977) Regulation by calcium ions of pyruvate carboxylation, pyruvate transport, and adenine nucleotide transport in isolated liver mitochondria. J Biol Chem 252:5372–5380

Friedmann N, Rasmussen H (1970) Calcium, manganese and hepatic gluconeogenesis. Biochim Biophys Acta 222:41–52

Gibson GE, Peterson C (1981) Aging decreases oxidative metabolism and the release and synthesis of acetylcholine. J Neurochem 37:978–984

Gibson GE, Jope R, Blass JP (1975) Decreased synthesis of acetylcholine accompanying impaired oxidation of pyruvic acid in rat brain minces. Biochem J 148:17–23

Giorgio Jr NA, Yip AT, Fleming J, Plaut GWE (1970) Diphosphoryridine nucleotide-linked isocitrate dehydrogenase from bovine heart. Polymeric forms and subunits. J Biol Chem 245:5469–5477

Goldstone TP, Crompton M (1982) Evidence for β-adrenergic activation of Na^+-dependent efflux of Ca^{2+} from isolated liver mitochondria. Biochem J 204:369–371

Goldstone TP, Duddridge RJ, Crompton M (1983) The activation of Na^+-dependent efflux of Ca^{2+} from liver mitochondria by glucagon and β-adrenergic agonists. Biochem J 210:463–472

Gupta BL, Hall TA (1978) Electron microprobe X-ray analysis of calcium. Ann NY Acad Sci 307:28–51

Hansford RG (1972) The effect of adenine nucleotides upon the 2-oxoglutarate dehydrogenase of blowfly flight muscle. FEBS Lett 21:139–141

Hansford RG (1980) Control of mitochondrial substrate oxidation. Curr Top Bioenerg 10:217–278

Hansford RG (1981) Effect of micromolar concentrations of free Ca^{2+} ions on pyruvate dehydrogenase interconversion in intact rat heart mitochondria. Biochem J 194:721–732

Hansford RG, Castro F (1981) Effect of micromolar concentrations of free calcium ions on the reduction of heart mitochondrial NAD(P) by 2-oxoglutarate. Biochem J 198:525–533

Hansford RG, Castro F (1982) Intramitochondrial and extramitochondrial free calcium ion concentrations of suspensions of heart mitochondria with very low, plausibly physiological, contents of total calcium. J Bioenerg Biomembr 14:361–376

Hansford RG, Chappell JB (1967) The effect of Ca^{2+} on the oxidation of glycerol phosphate by blowfly flight-muscle mitochondria. Biochem Biophys Res Commun 27:686–692

Hansford RG, Cohen L (1978) Relative importance of pyruvate dehydrogenase interconversion and feed-back inhibition in the effect of fatty acids on pyruvate oxidation by rat heart mitochondria. Arch Biochem Biophys 191:65–81

Heaton GM, Nicholls DG (1976) The calcium conductance of the inner membrane of rat liver mitochondria and the determination of the calcium electrochemical gradient. Biochem J 156:635–646

Hems DA, McCormack JG, Denton RM (1978) Activation of pyruvate dehydrogenase in the perfused rat liver by vasopressin. Biochem J 176:627–629

Henry PD, Shuchleib R, Davis J, Weiss ES, Sobel BE (1977) Myocardial contracture and accumulation of mitochondrial calcium in ischemic rabbit heart. Am J Physiol 233:H677–H684

Hiltunen JK, Hassinen IE (1976) Energy-linked regulation of glucose and pyruvate oxidation in isolated perfused rat heart. Role of pyruvate dehydrogenase. Biochim Biophys Acta 440:377–390

Hiraoka T, De Buysere M, Olson MS (1980) Studies of the effects of β-adrenergic agonists on the regulation of pyruvate dehydrogenase in the perfused rat heart. J Biol Chem 255:7604–7609

Hughes BP, Barritt GJ (1978) Effects of glucagon and $N^6 O^2$-dibutyryladenosine 3′:5′-cyclic monophosphate on calcium transport in isolated rat liver mitochondria. Biochem J 176:295–304

Hughes WA, Denton RM (1976) Incorporation of ^{32}Pi into pyruvate dehydrogenase phosphate in mitochondria from control and insulin-treated adipose tissue. Nature 264:471–473

Hunter DR, Komai H, Haworth RA, Jackson MD, Berkoff HA (1980) Comparison of Ca^{2+}, Sr^{2+}, and Mn^{2+} fluxes in mitochondria of the perfused rat heart. Circ Res 47: 721–727

Illingworth JA, Mullings R (1976) Pyruvate dehydrogenase activation after an increase in cardiac output. Biochem Soc Trans 4:291–292

Illingworth JA, Ford WCL, Kobayashi K, Williamson JR (1975) Regulation of myocardial energy metabolism. In: Roy P-E, Harris P (eds) Recent advances in studies on cardiac structure and metabolism, vol 8. University Park Press, Baltimore, pp 271–290

Jansson S-E, Gripenberg J, Härkönen M, Korpijoki P (1977) Methodological studies on the uptake of radiocalcium by nerve endings isolated from rat brain. Life Sci 20: 1431–1440

Jarett L, Seals JR (1979) Pyruvate dehydrogenase activation in adipocyte mitochondria by an insulin-generated mediator from muscle. Science 206:1407–1408

Johnson RN, Hansford RG (1975) The control of tricarboxylate-cycle oxidations in blowfly flight muscle. The steady-state concentrations of citrate, isocitrate, 2-oxoglutarate and malate in flight muscle and isolated mitochondria. Biochem J 146: 527–535

Johnson RN, Hansford RG (1977) The nature of controlled respiration and its relationship to protonmotive force and proton conductance in blowfly flight-muscle mitochondria. Biochem J 164:305–322

Joseph SK, Williamson JR (1983) The origin, quantitation, and kinetics of intracellular calcium mobilization by vasopressin and phenylephrine in hepatocytes. J Biol Chem 258:10425–10432

Joseph SK, Coll KE, Cooper RH, Marks JS, Williamson JR (1983) Mechanisms underlying calcium homeostasis in isolated hepatocytes. J Biol Chem 258:731–741

Jungas RL (1970) Effect of insulin on fatty acid synthesis from pyruvate, lactate, or endogenous sources in adipose tissue: evidence for the hormonal regulation of pyruvate dehydrogenase. Endocrinology 86:1368–1375

Katz AM (1970) Contractile proteins of the heart. Physiol Rev 50:63–158

Katz B, Miledi R (1967a) A study of synaptic transmission in the absence of nerve impulses. J Physiol 192:407–436

Katz B, Miledi R (1967b) The release of acetylcholine from nerve endings by graded electric pulses. Proc R Soc Lond (Biol) 167:23–38

Kauppinen RA, Hiltunen JK, Hassinen IE (1984) Mitochondrial membrane potential, transmembrane difference in the NAD^+ redox potential and the equilibrium of the glutamate-aspartate translocase in the isolated perfused rat heart. Biochim Biophys Acta (to be published)

Keilin D (1925) On cytochrome, a respiratory pigment, common to animals, yeast, and higher plants. Proc R Soc Lond (Biol) 98:312–339

Kessar P, Crompton M (1981) The α-adrenergic-mediated activation of Ca^{2+} influx into cardiac mitochondria. A possible mechanism for the regulation of intramitochondrial free Ca^{2+}. Biochem J 200: 379–388

Kiechle FL, Jarett L, Kotagal N, Popp DA (1981) Partial purification from rat adipocyte plasma membranes of a chemical mediator which simulates the action of insulin on pyruvate dehydrogenase. J Biol Chem 256:2945–2951

Klingenberg M, Buchholz M (1970) Localization of the glycerol-phosphate dehydrogenase in the outer phase of the mitochondrial inner membrane. Eur J Biochem 13:247–252

Langer GA, Nudd LM (1980) Addition and kinetic characterization of mitochondrial calcium in myocardial tissue culture. Am J Physiol 239:H769–H774

Lawlis VB, Roche TE (1980) Effect of micromolar Ca^{2+} on NADH inhibition of bovine kidney α-ketoglutarate dehydrogenase complex and possible role of Ca^{2+} in signal amplification. Mol Cell Biochem 32:147–152

Lawlis VB, Roche TE (1981a) Regulation of bovine kidney α-ketoglutarate dehydrogenase complex by calcium ion and adenine nucletides. Effects on $S_{0.5}$ for α-ketoglutarate. Biochemistry 20:2512–2518

Lawlis VB, Roche TE (1981b) Inhibition of bovine kidney α-ketoglutarate dehydrogenase complex by reduced nicotinamide adenine dinucleotide in the presence or absence of calcium ion and effect of adenosine 5′-diphosphate on reduced nicotinamide adenine dinucleotide inhibition. Biochemistry 20:2519–2524

Lee CO, Fozzard HA (1975) Activities of potassium and sodium ions in rabbit heart muscle. J Gen Physiol 65:695–708

Lee CO, Uhm DY, Dresdner K (1980) Sodium-calcium exchange in rabbit heart muscle cells: direct measurement of sarcoplasmic Ca^{2+} activity. Science 209:699–701

Llinás RR, Heuser JE (1977) Depolarization-release coupling systems in neurons. Neurosci Res Program Bull 15:556–687

Llinás R, Blinks JR, Nicholson C (1972) Calcium transient in presynaptic terminal of squid giant synapse: detection with aequorin. Science 176:1127–1129

Malmström K, Carafoli E (1976) The effect of Ca^{2+} on the oxidation of β-hydroxybutyric acid by heart mitochondria. Biochem Biophys Res Commun 69:658–664

Marban E, Rink TJ, Tsien RW, Tsien RY (1980) Free calcium in heart muscle at rest and during contraction measured with Ca^{2+}-sensitive microelectrodes. Nature 286: 845–850

Marshall SE, McCormack JG, Denton RM (1984) Role of Ca^{2+} ions in the regulation of intramitochondrial metabolism in rat epididymal adipose tissue. Evidence against a role for Ca^{2+} in the activation of pyruvate dehydrogenase by insulin. Biochem J 218:249–260

McCormack JG, Denton RM (1979) The effects of calcium ions and adenine nucleotides on the activity of pig heart 2-oxoglutarate dehydrogenase complex. Biochem J 180:533–544

McCormack JG, Denton RM (1980) Role of calcium ions in the regulation of intramitochondrial metabolism. Properties of the Ca^{2+} sensitive dehydrogenases within intact uncoupled mitochondria from the white and brown adipose tissue of the rat. Biochem J 190:95–105

McCormack JG, Denton RM (1981a) The activation of pyruvate dehydrogenase in the perfused rat heart by adrenaline and other inotropic agents. Biochem J 194:639–643

McCormack JG, Denton RM (1981b) A comparative study of the regulation by Ca^{2+} of the activities of the 2-oxoglutarate dehydrogenase complex and NAD^+-isocitrate dehydrogenase from a variety of sources. Biochem J 196:619–624

McCormack JG, Denton RM (1984) Role of Ca^{2+} ions in the regulation of intramitochondrial metabolism in rat heart. Evidence from studies with isolated mitochondria that adrenaline activates the pyruvate and 2-oxoglutarate dehydrogenase complexes by increasing the intramitochondrial concentration of Ca^{2+}. Biochem J 218:235–247

McCormack JG, England PJ (1983) Ruthenium red inhibits the activation of pyruvate dehydrogenase caused by positive inotropic agents in the perfused rat heart. Biochem J 214:581–585

Meijer AJ, van Woerkom GM, Steinman R, Williamson JR (1981) Inhibition by Ca^{2+} of carbamoylphosphate synthetase (ammonia). J Biol Chem 256:3443–3446

Mitchell P (1966) Chemiosmotic coupling in oxidative and photosynthetic phosphorylation. Glynn Research, Bodmin

Mitchell P (1979) Compartmentation and communication in living systems. Ligand conduction: a general catalytic principle in chemical, osmotic and chemiosmotic reaction systems. Eur J Biochem 95:1–20

Moore AL, Bonner Jr WD (1977) The effect of calcium on the respiratory responses of mung bean mitochondria. Biochim Biophys Acta 460:455–466

Moore CL (1971) Specific inhibition of mitochondrial Ca^{2+} transport by ruthenium red. Biochem Biophys Res Commun 42:298–305

Morad M, Goldman Y (1973) Excitation-contraction coupling in heart muscle: membrane control of development of tension. Prog Biophys Mol Biol 27:257–313

Morgan DG, Routtenberg A (1981) Brain pyruvate dehydrogenase: phosphorylation and enzyme activity altered by a training experience. Science 214:470–471

Morgan NG, Blackmore PF, Exton JH (1983) Modulation of the α_1-adrenergic control of hepatocyte calcium redistribution by increases in cyclic AMP. J Biol Chem 258: 5110–5116

Moriarity CM (1980) Kinetic analysis of calcium distribution in rat anterior pituitary slices. Am J Physiol 238:E167–E173

Murphy E, Mandel LJ (1982) Cytosolic free calcium levels in rabbit proximal kidney tubules. Am J Physiol 242:C124–128

Murphy E, Coll K, Rich TL, Williamson JR (1980) Hormonal effects on calcium homeostasis in isolated hepatocytes. J Biol Chem 255:6600–6608

Neely JR, Rovetto MJ, Oram JF (1972) Myocardial utilization of carbohydrate and lipids. Prog Cardiovasc Dis 15:289–329

Nicholls DG (1978) The regulation of extramitochondrial free calcium ion concentrations by rat liver mitochondria. Biochem J 176:463–474

Nicholls D, Åkerman K (1982) Mitochondrial calcium transport. Biochim Biophys Acta 683:57–88

Nicholls DG, Crompton M (1980) Mitochondrial calcium transport. FEBS Lett 111: 261–268

Nicholls DG, Scott ID (1980) The regulation of brain mitochondrial calcium-ion transport, the role of ATP in the discrimination between kinetic and membrane-potential calcium-ion efflux mechanisms. Biochem J 186:833–839

Nishiki K, Erecińska M, Wilson DF (1978) Energy relationship between cytosolic metabolism and mitochondrial respiration in rat heart. Am J Physiol 234:C73–C81

Otto DA, Ontko JA (1974) Regulation of the mitochondrial oxidation–reduction state in intact hepatocytes by calcium ions. Biochem Biophys Res Commun 61: 693–700

Otto DA, Ontko JA (1978) Activation of mitochondrial fatty acid oxidation by calcium. Conversion to the energized state. J Biol Chem 253:789–799

Ozawa E (1972) Activation of muscular phosphorylase *b* kinase by a minute amount of Ca ion. J Biochem (Tokyo) 71:321–331

Papa S, Tager JM, Francavilla A, de Haan EJ, Quagliariello E (1967) Control of glutamate dehydrogenase activity during glutamate oxidation in isolated rat-liver mitochondria. Biochim Biophys Acta 131:14–28

Parducz A, Joó F (1976) Visualization of stimulated nerve endings by preferential calcium accumulation of mitochondria. J Cell Biol 69:513–517

Patel TB, Olson MS (1982) Evidence for the regulation of the branched chain α-ketoacid dehydrogenase multienzyme complex by a phosphorylation/dephosphorylation mechanism. Biochemistry 21:4259–4265

Patzelt C, Löffler G, Wieland OH (1973) Interconversion of pyruvate dehydrogenase in the isolated perfused rat liver. Eur J Biochem 33:117–122

Paxton R, Harris RA (1982) Isolation of rabbit liver branched chain α-ketoacid dehydrogenase and regulation by phosphorylation. J Biol Chem 257:14433–14439

Pearse FJ, Walajtys-Rode E, Williamson JR (1980) Effects of work and acidosis on pyruvate dehydrogenase activity in perfused rat hearts. J Mol Bell Cardiol 12:499–510

Pettit FH, Roche TE, Reed LJ (1972) Function of calcium ions in pyruvate dehydrogenase phosphatase activity. Biochem Biophys Res Commun 49:563–571

Poggioli J, Berthon B, Claret M (1980) Calcium movements in "in situ" mitochondria following activation of α-adrenergic receptors in rat liver cells. FEBS Lett 115: 243–246

Randle PJ, England PJ, Denton RM (1970) Control of the tricarboxylate cycle and its interactions with glycolysis during acetate utilization in rat heart. Biochem J 117: 677–695

Randle PJ, Denton RM, Pask HT, Severson D (1974) Calcium ions and the regulation of pyruvate dehydrogenase. Biochem Soc. Symp 39:75–87

Rasmussen H, Waisman DM (1983) Modulation of cell function in the calcium messenger system. Rev Physiol Biochem Pharmacol 95:111–148

Reed LJ (1981) Regulation of mammalian pyruvate dehydrogenase complex by a phosphorylation-dephosphorylation cycle. Curr Top Cell Regul 18:95–106

Reed PW, Lardy HA (1972) A23187: a divalent cation ionophore. J Biol Chem 247: 6970–6977

Reinauer H, Muller-Ruchholtz ER (1976) Regulation of the pyruvate dehydrogenase activity in the isolated perfused heart of guinea pigs. Biochim Biophys Acta 444: 33–42

Reinhart PH, Taylor WM, Bygrave FL (1980) Trifluoperazine, an inhibitor of calmodulin action, antagonizes phenylephrine-induced metabolic responses and mitochondrial calcium fluxes in liver. FEBS Lett 120:71–74

Reinhart PH, Taylor WM, Bygrave FL (1982a) Studies on α-adrenergic-induced respiration and glycogenolysis in perfused rat liver. J Biol Chem 257:1906–1912

Reinhart PH, Taylor WM, Bygrave FL (1982b) Calcium ion fluxes induced by the action of α-adrenergic agonists in perfused rat liver. Biochem J 208:619–630

Robertson SP, Potter JD, Rouslin W (1982) The Ca^{2+} and Mg^{2+} dependence of Ca^{2+} uptake and respiratory function of porcine heart mitochondria. J Biol Chem 257: 1743–1748

Rolleston FS (1972) A theoretical background to the use of measured concentrations of intermediates in study of the control of intermediary metabolism. Curr Top Cell Regul 5:47–75

Sacktor B (1975) Regulation of intermediary metabolism with special reference to the control mechanisms in insect flight muscle. Adv Insect Physiol 7:267–347

Scarpa A, Graziotti P (1973) Mechanisms for intracellular calcium regulation in heart. I. Stopped-flow measurements of Ca^{++} uptake by cardiac mitochondria. J Gen Physiol 62:756–772

Schweitzer ES, Blaustein MP (1980) Calcium buffering in presynaptic nerve terminals. Free calcium levels measured with Arsenazo III. Biochim Biophys Acta 600:912–921

Scott ID, Akerman KEO, Nicholls DG (1980) Calcium ion transport by intact synaptosomes. Intrasynaptosomal compartmentation and the role of the mitochondrial membrane potential. Biochem J 192:873–880

Seals JR, Czech MP (1980) Evidence that insulin activates an intrinsic plasma membrane protease in generating a secondary chemical mediator. J Biol Chem 255: 6529–6531

Severson DL, Denton RM, Pask HT, Randle PJ (1974) Calcium and magnesium ions as effectors of adipose-tissue pyruvate dehydrogenase phosphate phosphatase. Biochem J 140:225–237

Severson DL, Denton RM, Bridges BJ, Randle PJ (1976) Exchangeable and total calcium pools in mitochondria of rat epididymal fat pads and isolated fat cells. Role in the regulation of pyruvate dehydrogenase activity. Biochem J 154:209–223

Sies H, Graf P, Crane D (1983) Decreased flux through pyruvate dehydrogenase during calcium ion movements induced by vasopressin, α-adrenergic agonists and the ionophore A23187 in perfused rat liver. Biochem J 212:271–278

Siess EA, Brocks DG, Wieland OH (1978) Comparative studies on the influence of hormones on metabolite compartmentation in isolated liver cells during gluconeogenesis from lactate. Biochem Soc Trans 6:1139–1144

Smith CM, Bryła J, Williamson JR (1974) Regulation of mitochondrial α-ketoglutarate metabolism by product inhibition at α-ketoglutarate dehydrogenase. J Biol Chem 249:1497–1505

Somlyo AP, Somlyo AV, Shuman H, Sloane B, Scarpa A (1978) Electron probe analysis of calcium compartments in cryo sections of smooth and striated muscles. Ann NY Acad Sci 307:523–544

Somlyo AP, Somlyo AV, Shuman H (1979) Electron probe analysis of vascular smooth muscle. J Cell Biol 81:316–335

Somlyo AP, Somlyo AV, Shuman H, Scarpa A, Endo M, Inesi G (1981) Mitochondria do not accumulate significant Ca concentrations in normal cells. In: Bronner F, Peterlik M (eds) Calcium and phosphate transport across biomembranes. Academic, New York, pp 87–93

Stucki JW, Ineichen A (1974) Energy dissipation by calcium recycling and the efficiency of calcium transport in rat liver mitochondria. Eur J Biochem 48:365–375

Studer RK, Borle AB (1979) Effect of pH on the calcium metabolism of isolated rat kidney cells. J Membr Biol 48:325–341

Sugano T, Shiota M, Khono H, Shimada M, Oshino N (1980) Effects of calcium ions on the activation of gluconeogenesis by norepinephrine in perfused rat liver. J Biochem 87:465–472

Sugden MC, Ball AJ, Ilic V, Williamson DH (1980b) Stimulation of [1-^{14}C] oleate oxidation to $^{14}CO_2$ in isolated rat hepatocytes by vasopressin: effects of Ca^{2+}. FEBS Lett 116:37–40

Sugden MC, Ball AJ, Williamson DH (1980a) A second site of vasopressin action on [1-^{14}C] oleate metabolism in isolated rat hepatocytes: Increased formation of $^{14}CO_2$. Biochem Soc Trans 8:591–592

Taylor WM, Prpić V, Exton JH, Bygrave FL (1980) Stable changes to calcium fluxes in mitochondria isolated from rat liver perfused with α-adrenergic agonists and with glucagon. Biochem J 188:443–450

Taylor WM, Reinhart PH, Bygrave FL (1983) Stimulation by α-adrenergic agonists of Ca^{2+} fluxes, mitochondrial oxidation and gluconeogenesis in perfused rat liver. Biochem J 212:555–565

Tsien RY, Pozzan T, Rink TJ (1982) Calcium homeostasis in intact lymphocytes: cytoplasmic free calcium monitored with a new, intracellularly trapped fluorescent indicator. J Cell Biol 94:325–334

Vághy PL, Johnson JD, Matlib MA, Wang T, Schwartz A (1982) Selective inhibition of Na^{+}-induced Ca^{2+} release from heart mitochondria by Diltiazem and certain other Ca^{2+} antagonist drugs. J Biol Chem 257:6000–6002

van der Meer R, Akerboom TPM, Groen AK, Tager JM (1978) Relationship between oxygen uptake of perifused rat-liver cells and the cytosolic phosphorylation state calculated from indicator metabolites and a redetermined equilibrium constant. Eur J Biochem 84:421–428

van Rossum GDV, Smith KP, Beeton P (1976) Role of mitochondria in control of calcium content of liver slices. Nature 260:335–337

Veloso D, Guynn RW, Oskarsson M, Veech RL (1973) The concentrations of free and bound magnesium in rat tissues. Relative constancy of free Mg^{2+} concentrations. J Biol Chem 248:4811–4819

Vinogrador A, Scarpa A (1973) The initial velocities of calcium uptake by rat liver mitochondria. J Biol Chem 248:5527–5531

Walajtys EI, Gottesman DP, Williamson JR (1974) Regulation of pyruvate dehydrogenase in rat liver mitochondria by phosphorylation–dephosphorylation. J Biol Chem 249:1857–1865

Weiss L, Löffler G, Schirmann A, Wieland O (1971) Control of pyruvate dehydrogenase interconversion in adipose tissue by insulin. FEBS Lett 15:229–231

Wendt-Gallitelli MF, Jacob R (1982) Rhythm-dependent role of different calcium stores in cardiac muscle: X-ray microanalysis. J Mol Cell Cardiol 14:487–492

Wernette ME, Ochs RS, Lardy HA (1981) Ca^{2+} stimulation of rat liver mitochondrial glycerophosphate dehydrogenase. J Biol Chem 256:12767–12771

Whiting JA, Barritt GJ (1982) On the mechanism by which hormones induce the release of Ca^{2+} from mitochondria in the liver cell. Biochem J 206:121–129

Wieland OH (1983) The mammalian pyruvate dehydrogenase complex: structure and regulation. Rev Physiol Biochem Pharmacol 96:123–170

Williamson JR (1975) Effects of epinephrine on glycogenolysis and myocardial contractility. Handb Physiol (Sect 7) 6:605–636

Williamson JR (1979) Mitochondrial function in the heart. Annu Rev Physiol 41:485–506

Williamson JR, Cooper RH (1980) Regulation of the citric acid cycle in mammalian systems. FEBS Lett 117:K73–K85

Williamson JR, Murphy E (1980) Calcium homeostasis and compartmentation in liver. In: Thurman RG (ed) Advances in experimental medicine and biology. Plenum, New York, pp 671–688

Williamson JR, Cooper RH, Hoek JB (1981) Role of calcium in the hormonal regulation of liver metabolism. Biochim Biophys Acta 639:243–295

Wilson DF, Owen CS, Holian A (1977) Control of mitochondrial respiration: a quantitative evaluation of the roles of cytochrome c and oxygen. Arch Biochem Biophys 182:749–762

Wonnacott S, Marchbanks RM, Fiol C (1978) Ca^{2+} uptake by synaptosomes and its effect on the inhibition of acetylcholine release by botulinum toxin. J Neurochem 30:1127–1134

Zoccarato F, Nicholls D (1982) The role of phosphate in the regulation of the independent calcium-efflux pathway of liver mitochondria. Eur J Biochem 127:333–338

Rev. Physiol. Biochem. Pharmacol., Vol. 102
© by Springer-Verlag 1985

Acetylcholine-Receptor-Mediated Ion Fluxes in *Electrophorus electricus* and *Torpedo californica* Membrane Vesicles

DEREK J. CASH[1], HITOSHI AOSHIMA[2], ELENA B. PASQUALE[3], and GEORGE P. HESS[4]

Contents

Fakultät für Biologie, Universität Konstanz, West Germany

1 Present address: Department of Biochemistry, University of Missouri-Columbia, Neurochemistry Unit, Missouri Institute of Psychiatry, St. Louis, MO 63139, USA

2 Present address: Department of Chemistry, Yamaguchi University, Yamaguchi 753, Japan

3 Present addresses: Department of Biology, University of California, San Diego, CA, USA, and University of Parma, Parma, Italy

4 Section of Biochemistry, Division of Biological Sciences, 270 Clark Hall, Cornell University, Ithaca, NY 14853, USA

1 Introduction

Because of its accessibility in neuromuscular junctions, and especially in the electric organ of electric fish, the acetylcholine receptor protein is studied as a general model for a number of receptors which form ion-conducting channels through cell membranes. (For recent reviews of the properties of the acetylcholine receptor see *Adams* 1981; *Karlin* 1980; *Hess* et al. 1983.) Here we focus our attention on the effect of acetylcholine concentration on the acetylcholine receptor-controlled movement of ions across membranes. The rate of this transmembrane ion flux determines the amplitude of the electrical signal and thereby whether the signal will be transmitted. The relationship between the concentration of the chemical signal and ion translocation rates may be particularly significant for an understanding of the way the nervous system functions: environmental changes are believed to affect the transmission of signals between cells. They are known to affect receptor-controlled processes in the sea snail (for a review see *Kandel* 1981).

Here we review our chemical kinetic measurements of the transmembrane flux of cations (K^+, $^{86}Rb^+$, Na^+) that is controlled by the acetylcholine receptor. The measurements were made mainly using membrane vesicles prepared from the electric organ (electroplax) of the electric eel (*Electrophorus electricus*). Some experiments were also done with vesicles prepared from the *Torpedo californica* electroplax. We will describe first the preparation and purification of membrane vesicles, and the kinetic techniques which allow ion translocation processes to be measured in the millisecond to second time region. We will then describe a minimum model that relates ligand-binding steps to ion translocation rates and the evaluation of all the rate and equilibrium constants pertaining to the model. We will show that this model, unlike all the other models previously proposed, quantitatively accounts for the ion translocation process over wide ranges of acetylcholine (5000-fold) and carbamoylcholine (2000-fold) concentrations. Finally, we will discuss the relationship between chemical kinetic measurements and the elegant statistical methods (*Katz* and *Miledi* 1970, 1972, 1977; *Neher* and *Sakmann* 1976; *Anderson* and *Stevens* 1973) which have been used to investigate elementary steps of the channel-opening process in muscle cells.

The bioelectric phenomena exhibited by the *E. electricus* electroplax, which is the source of our vesicle preparation, have been studied extensively by *Nachmansohn* (1955; *Nachmansohn* and *Neumann* 1975) at Columbia Universtiy during the past 40 years. Identification of compounds that alter the electrical properties of a single electroplax cell became possible when *Schoffeniels* and *Nachmansohn* (1957) succeeded in making electrical recordings with such cells. Chemical kinetic measurements with vesicles prepared from this organ became possible when *Kasai* and *Changeux* (1971a,b,c) developed a membrane preparation which contained vesicles that exhibited receptor function, as measured by the translocation of radioactive tracer cations in the presence of acetylcholine.

For many years the electroplax of *E. electricus* has been used in studies of bioelectric phenomena (*Keynes* and *Martins-Ferreira* 1953; *Nachmansohn* and *Neumann* 1975). *E. electricus* electroplax cells continue to be of interest for electrophysiological investigations of responses to acetylcholine in intact cells (*Schoffeniels* and *Nachmansohn* 1957; *Podleski* and *Bartels* 1963). The vesicles used in the experiments to be described here were made from plasma membranes of the same tissue.

The advantages of working with membrane vesicles, rather than with intact cells, have been pointed out in a number of reviews (*Lehninger* 1964; *Racker* 1970; *Kaback* 1970). The concentrations of solutions both inside and outside the vesicles can be controlled, internally by equilibrating the vesicles with defined buffer for an appropriate period of time, and externally by suspending and diluting the vesicles in defined buffers. The acetylcholine receptor protein can be specifically marked with α-bungarotoxin labeled with radioactive iodine (*Lee* 1972). α-Bungarotoxin is a snake neurotoxin which binds specifically and irreversibly to the acetylcholine receptor and inhibits the functioning of the protein (*Lee* 1972). The concentration of receptor sites can then be determined. Therefore, it seemed possible in these vesicles to investigate receptor-controlled flux through a determined number of receptors under well-defined experimental conditions.

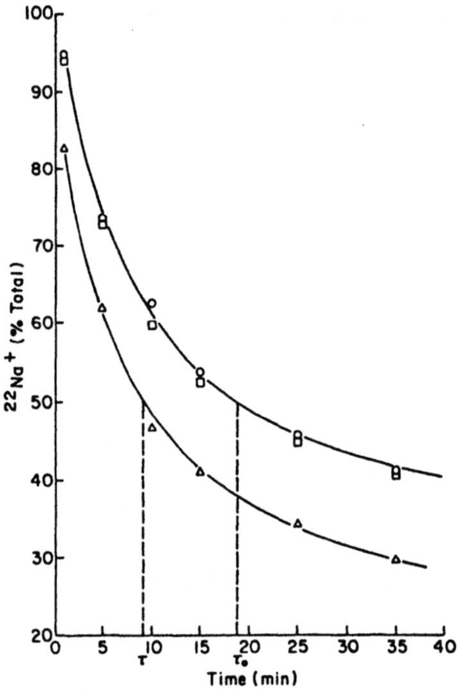

Fig. 1. Demonstration of receptor-controlled ion flux by the procedure of *Kasai* and *Changeux* (1971a). Vesicles were incubated with $^{22}Na^+$ overnight, and then diluted 70-fold with buffer alone (○,□) or with buffer containing 1×10^{-4} *M* carbamoylcholine (△). After various periods of time (indicated on the ***abscissa***), the vesicles were removed from the dilution buffer by Millipore filtration and their tracer content was determined. The time for half-equilibration of the vesicles with the dilution buffer is given for the control curve (τ_0) and for the carbamoylcholine-induced flux curve (τ). (*Hess* et al. 1975a)

2 Measurements of Tracer Ion Flux in Membrane Vesicles Prepared from *Electrophorus electricus* Electroplax

2.1 Experiments with Heterogeneous Vesicle Preparations

When the approaches developed (*Lehninger* 1964; *Bangham* et al. 1965; *Kaback* 1970; *Passow* 1969; *Racker* 1970) to study transmembrane processes in artificial and natural membrane vesicles were applied to *E. electricus* electroplax membranes, sealed vesicles were obtained (*Kasai* and *Changeux* 1971a,b,c). The vesicles were prepared by removing the electric organ, homogenizing the diced tissue, and by two centrifugation steps, the second one involving a sucrose gradient (*Kasai* and *Changeux* 1971a,b,c).

2.1.1 Attempts to Measure Acetylcholine-Receptor-Controlled Ion Flux

The vesicles were equilibrated with radioactive tracer ions (usually $^{22}Na^+$). Efflux of the tracer ions was then measured by diluting the vesicles with a buffer and, after various periods of time, separating the vesicles from the suspension by Millipore filtration. The amount of radioactive tracer ions remaining in the vesicles was then determined. This technique demonstrated an increased permeability of the vesicles to Na^+ and K^+ in the presence of acetylcholine or carbamoylcholine (*Kasai* and *Changeux*

1971a,b,c). $^{22}Na^+$ efflux data obtained in similar experiments are shown in Fig. 1. The flux was measured in a vesicle preparation in the absence and presence of carbamoylcholine, an acetylcholine analog that is not hydrolized by acetylcholinesterase.

Pharmacological similarities between ligand-induced ion flux in the vesicles and the electrophysiological response of electroplax cells were reported (*Kasai* and *Changeux* 1971a,b,c). Receptor-controlled fluxes of Na^+, K^+, and Ca^{2+} were demonstrated; fluxes of larger ions, choline and tetramethylammonium, were not detected under these conditions. As in the intact electroplax cells, receptor activity could be abolished by compounds such as *d*-tubocurarine and snake neurotoxins (for instance, α-bungarotoxin) (*Kasai* and *Changeux* 1971a,b,c). Subsequently, ion efflux was measured in both *E. electricus* and *Torpedo* spp. membrane vesicles in several laboratories (*McNamee* and *McConnell* 1973; *Hazelbauer* and *Changeux* 1974; *Sugiyama* et al. 1975; *Popot* et al. 1976; *Hess* et al. 1975a, 1976; *Miller* et al. 1978).

When the rate of receptor-controlled transfer of ions across membranes was calculated from measurements made with vesicles, the results were lower, by a factor of 10^5, then those determined by electrophysiological techniques in intact cells (*Katz* and *Miledi* 1970, 1972; *Neher* and *Stevens* 1977). In addition, receptor inactivation (desensitization) induced by acetylcholine that was observed in electrophysiological experiments with cells (*Katz* and *Thesleff* 1957) was not observed in efflux measurements made in *E. electricus* vesicles (*Kasai* and *Changeux* 1971a,b,c). These discrepancies made the relevance of experiments performed with vesicles doubtful.

2.1.2 Problems Encountered in Measurements of Receptor-Controlled Flux in Heterogeneous Vesicle Populations

Because the potential utility of measurements with vesicles was demonstrated with other systems (*Bangham* et al. 1965; *Lehninger* 1964; *Racker* 1970; *Kaback* 1970), we determined some of the reasons for the discrepancies between measurements in vesicles and in cells. We were then able to develop methods that avoided the difficulties inherent in the measurement of acetylcholine-receptor-controlled ion flux in vesicles (*Hess* et al. 1975a, 1979; *Hess* and *Andrews* 1977; *Kim* and *Hess* 1981).

The major factor that contributes to the low rates of receptor-controlled efflux of tracer ions that were measured is the presence of leaky vesicles that contain the bulk of the exchangeable tracer ions (80%–90%) but that do not respond to acetylcholine (*Hess* et al. 1975a). By a process which we shall explain, the loss of ions from leaky vesicles obscures the receptor-controlled flux of cations from a small population of vesicles that contain

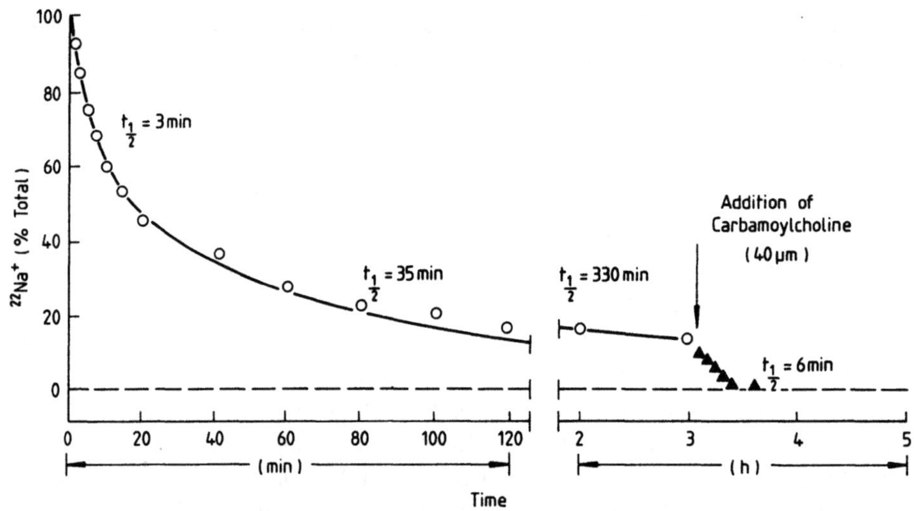

Fig. 2. Kinetic analysis of $^{22}Na^+$ efflux from vesicles (pH 7.0; 4°C). Experimental conditions as in Fig. 1. The observed time course of the efflux in the absence of carbamoylcholine is fitted to the sum of three exponentials:

$$^{22}Na_t^+ = {}^{22}Na_{t=0}\,(\alpha e^{-k_1 t} + \beta e^{-k_2 t} + \gamma e^{-k_3 t})$$

The half-times of the three processes are shown ($t_{1/2}$). Carbamoylcholine-induced efflux was initiated at the time indicated by the arrow. (*Hess* and *Andrews* 1977)

only 10%–20% of the exchangeable tracer ions (*Hess* et al. 1975a). The inactivation (desensitization) of the receptor by its own ligand had occurred before the first time-point in these measurements. Receptor desensitization can be observed in these vesicles (*Hess* et al. 1978; *Epstein* et al. 1980) and the rate of the process can be analyzed quantitatively (*Aoshima* et al. 1981).

The vesicle preparation of *Kasai* and *Changeux* (1971a) was found to contain three types of vesicle, each with a characteristic $t_{1/2}$ for efflux (3, 35, and 330 min) (*Hess* et al. 1975a; *Hess* and *Andrews* 1977) (Fig. 2).

Sensitivity to acetylcholine or carbamoylcholine is exhibited by the vesicles that are rather impermeable to inorganic ions ($t_{1/2}$ = 330 min) and that contain between 10% and 20% of the tracer ion content of the total preparation (*Hess* and *Andrews* 1977) (Fig. 2). When tracer ions in vesicles which do not contain functional receptors are allowed to equilibrate with outside buffer (Fig. 2) and then carbamoylcholine is added, efflux follows a single exponential rate law (*Hess* et al. 1975a) (Fig. 3).

How is the rate of the carbamoylcholine-sensitive flux obscured? Let us assume that the carbamoylcholine-induced $^{22}Na^+$ efflux, involving about 10% of the tracer content of the vesicle population, occurs within 0.5 s of the addition of carbamoylcholine. The $t_{1/2}$ value for the efflux of tracer ions from the total vesicle population is observed when 50% of the tracer content of the vesicle population has been lost. Even in the presence of

Fig. 3. First order plot of $^{22}Na^+$ efflux due to the presence of 0.2 m*M* carbamoylcholine. The *ordinate* gives $^{22}Na^+_t$, the concentration of radioactive ions inside the vesicles at time t (pH 7.0; 4°C). (*Hess* 1979)

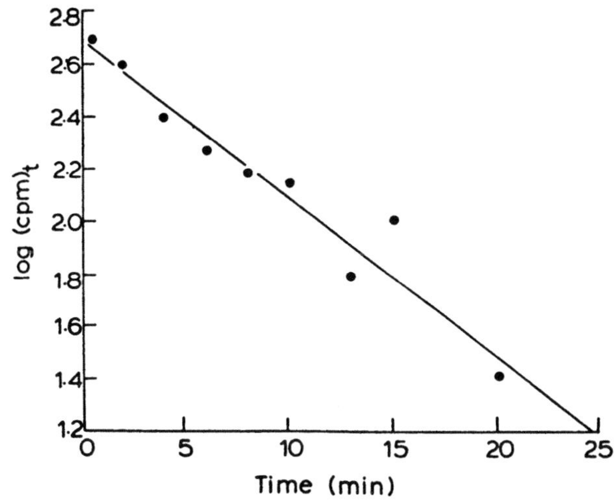

carbamoylcholine, this 50% depletion of tracer ions from the vesicle population is mainly due to efflux from vesicles without receptors. The observed difference in $t_{1/2}$ between the experiments in the absence of and in the presence of an activating ligand in the example given (τ_0 and τ, respectively in Fig. 1) is due to the difference in the time in which the vesicles without receptors lose an amount of tracer ions corresponding to 50% (in the absence of carbamoylcholine) and 40% (in the presence of carbamoylcholine) of the tracer ion content of the vesicle preparation. This is a process which occurs in minutes, as can be seen in Fig. 1. In the presence of an activating ligand, 50% of the tracer ion is lost in about 9 min. Now suppose that the receptors are inactivated or inhibited and receptor-controlled flux occurs 100 times more slowly, that is, the vesicles with inactivated receptors are depleted of tracer ions in 50 s, rather than in 0.5 s. This would increase by only 10% the observed time in which 50% of the tracer ions are lost in the presence of ligand. It can easily be seen why the inactivation process was not observed under the conditions employed by *Kasai* and *Changeux* (1971a).

In general, the $t_{1/2}$ value of the efflux from a heterogeneous vesicle population in the presence of acetylcholine depends to only a minor extent on the receptor-controlled flux, which is very fast compared with the passive flux (*Hess* et al. 1979; *Cash* and *Hess* 1980). Instead, the $t_{1/2}$ value depends heavily on the $t_{1/2}$ value of the passive efflux and on the relative amounts of tracer ions in the vesicle populations containing active and inactive receptors. Recently the effects of vesicle heterogeneity and other aspects of ion flux measurements have been treated theoretically by *Neumann* and his colleagues (*Bernhardt* and *Neumann* 1978, 1980).

2.1.3 Other Problems Encountered in Ion Flux Measurements

When tracer ion efflux induced by carbamoylcholine or gramicidin is measured after incubation of the vesicles with radioactive ions for 15 min or longer, 25%–30% of the tracer ion content of the vesicles is lost slowly with a $t_{1/2}$ value of 3 min, even in the presence of high concentrations of carbamoylcholine or gramicidin. This means that a fraction of the tracer ions becomes trapped in the vesicles and its subsequent efflux is limited by factors other than the receptor (*Kim* and *Hess* 1981). This problem can be avoided, providing one restricts the time in which the flux rate (*Kim* and *Hess* 1981) after the addition of activating ligand is measured. In practice we aim for conditions which allow measurements to be completed within 30 s.

For reasons that we shall describe later, at low ligand concentrations such as were used in the early experiments or at low ratios of receptor concentration to the internal volume of the vesicles, the observed flux is due to desensitized receptor, and the apparent receptor:ligand dissociation constants evaluated from such measurements are characteristic of the inactivated (desensitized) receptor (see Sect. 6).

The state of aggregation of the vesicles affects the amount of radioactivity retained on the Millipore filter. When one dilutes vesicles equilibrated with tracer ions about 100-fold and then removes the vesicles from suspension by filtration (final vesicle concentration 80 μg membrane protein/ml, 1 ml filtered) and determines the tracer ion content of the vesicles on the filter, an apparent reduction in the content is observed. The $t_{1/2}$ value of this process is about 3 min. Simultaneous light-scattering measurements indicated that disaggregation of the vesicles, with a $t_{1/2}$ value of 3 min, occurs when the vesicles are diluted 100-fold (*Kim* and *Hess* 1981). This problem can be avoided if the vesicles are diluted less than tenfold, to a final concentration of 80 μg membrane protein/ml.

Three different approaches were used to measure receptor-controlled flux in the absence of flux from leaky vesicles. The first was to allow the leaky vesicles containing tracer ions to equilibrate with nonradioactive ions in the external buffer before adding carbamoylcholine (*Hess* et al. 1975a) (Fig. 2). The second approach was physically to separate the vesicles (*Hess* and *Andrews* 1977; *Sachs* et al. 1982). The third approach was to measure the rapid receptor-controlled flux before the slower passive flux in the vesicle population became important. The techniques used are described in greater detail in the following sections.

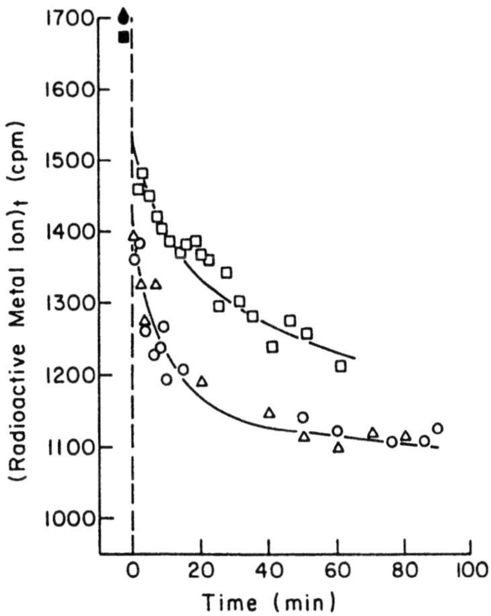

Fig. 4. $^{22}Na^+$ and $^{86}Rb^+$ efflux as a function of time under physiological conditions. For the efflux measurements, sucrose-free solutions of vesicles were prepared by the procedure of *Fu* et al. (1974). The vesicles, resuspended in a potassium Ringer's solution (150 m*M* KCl, 9.4 m*M* NaCl, 3 m*M* $CaCl_2$, 1.5 m*M* $MgCl_2$, 1 m*M* phosphate at pH 7.0) were incubated overnight with $^{86}RbCl$ or $^{22}NaCl$ at 4°C. Efflux was initiated by dilution into a sodium eel Ringer's solution (169 m*M* NaCl, 5 m*M* KCl, 3 m*M* $CaCl_2$, 1 m*M* phosphate at pH 7.0) at 4°C. After 120 min, ligand was added. ○, $^{86}Rb^+$ efflux, 1 m*M* carbamoylcholine; △, $^{86}Rb^+$ efflux, 10 *μM* acetylcholine (vesicles were preincubated with 56 *μM* Tetram which inhibitis acetylcholinesterase); □, $^{22}Na^+$ efflux, 1 m*M* carbamoylcholine. *Solid symbols* represent the respective average of several experimental points taken before the addition of ligand. (*Hess* et al. 1978)

2.2 Kinetic Separation of Receptor-Controlled Flux from Flux Not Controlled by the Receptor

The receptor-controlled tracer efflux was measured after the vesicles not containing functional receptors had lost their tracer ion content (Fig. 2) (*Hess* et al. 1975a). This technique was used to obtain the data shown in Figs. 3 and 4. It can be seen that the carbamoylcholine-induced efflux of tracer ions follows a single exponential rate law when the concentration of carbamoylcholine is less than 0.05 m*M* (Fig. 3). At this ligand concentration the efflux is due to an equilibrium mixture of active and inactive receptor states (see Sect. 4). At higher concentrations of carbamoylcholine (>0.2 m*M*), the efflux curve becomes biphasic (Fig. 4) (*Hess* et al. 1978). The second, slower phase is flux considered to be due to the inactivated (desensitized) receptor. However, a different process which reflects a property of the vesicles with a $t_{1/2}$ value of 3 min becomes rate limiting when the vesicles are exposed to tracer ions for more than 30 s (*Kim* and

Fig. 5a

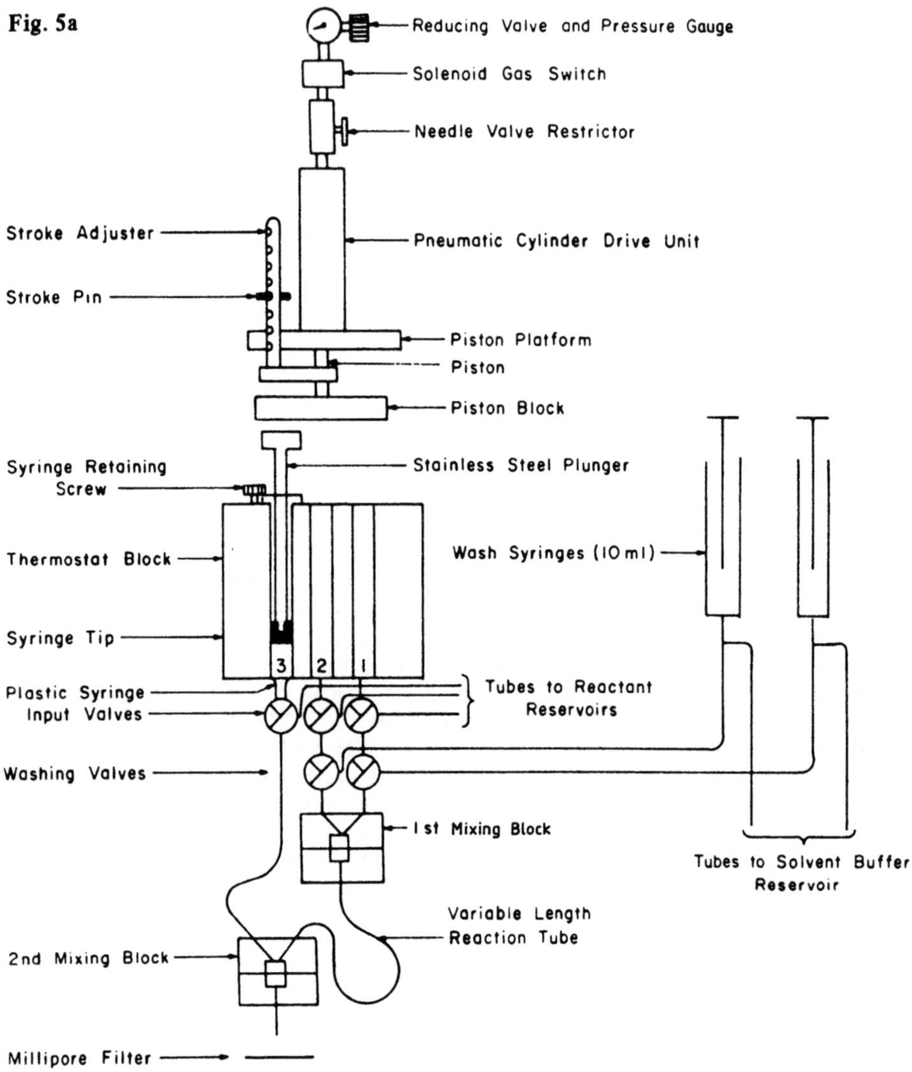

Fig. 5a,b. Pulsed quench flow apparatus used in measurements with *E. electricus* electroplax membrane vesicles. Comparison of continuous and pulsed mode quench flow. **a** Continuous quench flow. The reaction time, t, varies with the volume between initiation in the first mixer and quenching in the second mixer.

$$t + t_D = T\,(v/V)$$

V is the total volume displaced and T is the time for this displacement; v is the volume between initiation and quenching and t_D is the determined dead time. **b** Tubing arrangement for two pulsed incubations with three displacements. Reagents from syringes 1 and 2 are incubated in tube 1 before being displaced by solvent from syringes 4 and 5, mixed with reagent from syringe 6 and incubated in tube 2 before being displaced by solvent from syringes 7 and 8 and mixed with reagent from syringe 9.

In our measurements the solution containing the membranes and reagents is expelled onto a Millipore filter and washed with buffer. The tubing is made from Teflon. (*Cash* and *Hess* 1981)

Fig. 5b

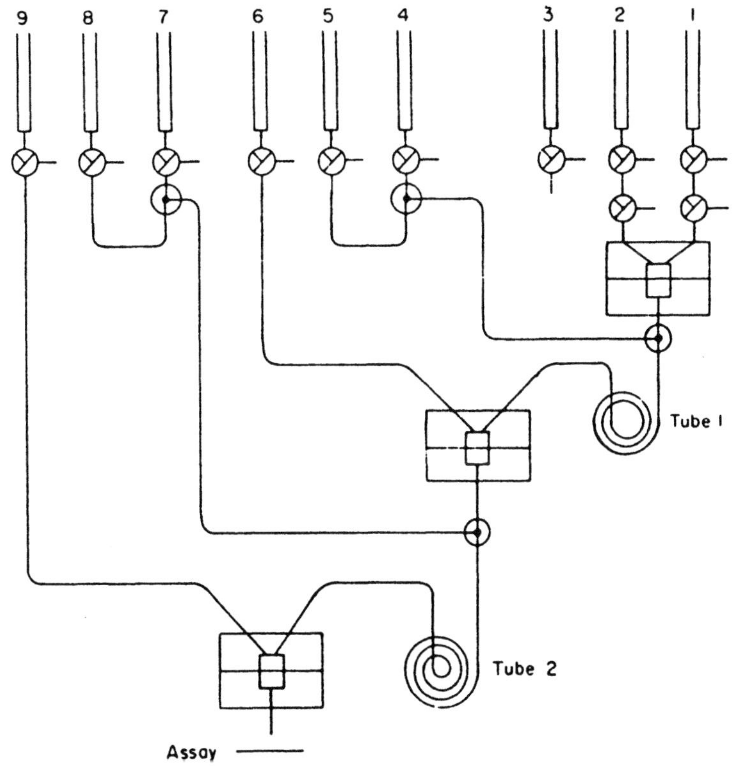

Hess 1981; *Cash* and *Hess* 1981). We measured influx to avoid exposure of the vesicles to tracer ions before the measurements were made.

2.3 Measurements of Receptor-Controlled Tracer Ion Flux in the Millisecond to Second Time Region Using Quench Flow Techniques

Techniques of rapid mixing of solutions have been used extensively with proteins to follow reactions occurring in the time range of milliseconds to seconds (*Roughton* and *Chance* 1963; *Chance* et al. 1964; *Hammes* 1978). Quench flow techniques have been applied to enzyme systems (*Fersht* and *Jakes* 1975; *Lymn* and *Taylor* 1970; *Froehlich* and *Taylor* 1976; *Martonosi* et al. 1974). Transmembrane ion transport has been demonstrated using quench flow technique in studies of ATP synthesis by submitochondrial particles (*Thayer* and *Hinkle* 1975) and calcium uptake by sarcoplasmic reticulum vesicles (*Kurzmack* et al. 1977; *Verjovski-Almeida* and *Inesi* 1979; *Sumida* et al. 1978). With a continuous quench flow system (Fig. 5a), the reaction time depends on the flow rate and the distance between an initiating and a quenching event. Thus, as the reaction time increases with increasing length of the reaction tube, the recovery of reac-

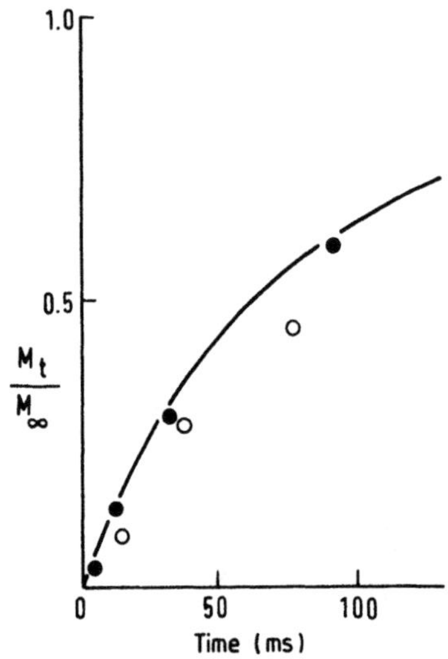

Fig. 6. Acetylcholine-receptor-mediated flux of $^{86}Rb^+$ into membrane vesicles at 1°C. Eel Ringer's solution, pH 7.0, 300 μM acetylcholine. Measurements in the time region 5–100 ms illustrate the time resolution of the technique. M_t and M_∞ represent the $^{86}Rb^+$ content of the vesicles at time t and at the end of the measurement, respectively. The *closed* and *open circles* represent data from two experiments. (*Cash* et al. 1981)

tant decreases, due to its increased retention in the tube. There is an upper limit to the tube length (reaction time) imposed by the quantity of material available for the investigations. This limitation was overcome by the pulsed quench flow system of *Fersht* (*Fersht* and *Jakes* 1975), in which the reaction liquid is stationary between the initiation and quenching, and the quantity of sample recovered is independent of the reaction time (Fig. 5b). The initiation and quenching events are controlled independently of each other, and the reaction time is preset with an electronic timer. With this technique there is a lower limit of reaction time equal to the syringe stroke time. The shortest possible reaction time is limited by the mixing of the solutions and is around a millisecond. In practice, the time resolution of the instrument is given by the precision of determination of the dead time. In measurements of acetylcholine-receptor-controlled ion flux in vesicles, a reaction time of 5 ms can be obtained precisely (Fig. 6).

When using a quench flow technique with vesicles, it is possible that breakage of the vesicles can occur, particularly with long lengths of tubing. This was investigated using iodosulfanilic acid [^{125}I]-labeled membrane as an internal standard while measuring the internal volume with $^{86}Rb^+$. The counts from ^{125}I (lower energy) and ^{86}Rb (higher energy radiation) in the same sample can be resolved. A decreased ratio of ^{86}Rb:^{125}I indicates breakage. Breakage is negligible under the conditions which were used. With the vesicle system, the measured mixing-quenching dead time was longer than that measured using a model hydrolysis reaction at the same flow rates and solution viscosity with a similar concentration excess of quench solution. In principle, besides the macroscopic physical properties of the mixer and solution, the mixing time as measured by a fast chemical reaction is a property of the chemical system and decreases with increasing diffusion coefficient of the reactants (*Toor* 1975). The increased

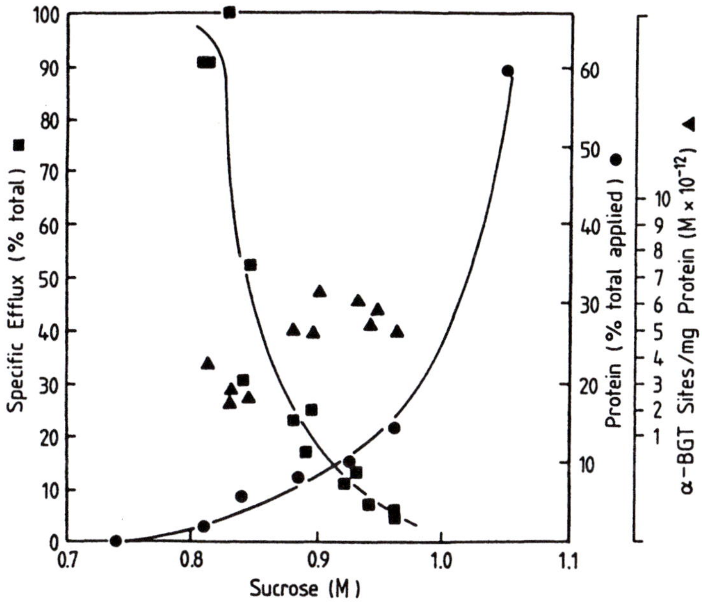

Fig. 7. Purification of vesicles that contain active receptors by sucrose-190 m*M* cesium chloride density gradient centrifugation. A population of vesicles rich in α-bungarotoxin (*α-BGT, solid triangles*) binding sites and of similar density was collected by centrifugation on a discontinuous sucrose gradient. The material between layers of 0.9 *M* and 1.2 *M* sucrose was collected. The vesicles were then equilibrated with 190 m*M* NaCl for 24 h, long enough for all vesicles to equilibrate with the outside solution (see Fig. 2). The 190 m*M* NaCl in the external solution was exchanged for 190 m*M* CsCl using a Sephadex G-25 column. The vesicles were then allowed to re-equilibrate for 2 h. According to the analysis of the efflux data in Fig. 2, this allows all the nonspecific vesicles to fill with CsCl ($t_{1/2}$ = 35 min), while the carbamoylcholine-sensitive vesicles still retain most of their NaCl content. The specific efflux is defined as $^{22}Na^+$ efflux per mg membrane protein in presence of 1 m*M* carbamoylcholine, and the total efflux as the sum of specific and nonspecific efflux per mg membrane protein. The measurement of the carbamoylcholine-dependent efflux of ions is described in the legend of Fig. 1. (*Hess* 1979). Most of the membrane protein (*solid circles*) was found in the denser portion of the gradient, that is, in the CsCl filled, nonspecific vesicles without functional receptors. *Solid squares* indicate the percentage of Na^+ inside the vesicles whose flux rate is affected by ligands such as carbamoylcholine. The 0.8 *M* sucrose fractions contained only those vesicles responding to carbamoylcholine

dead time observed with the vesicles is consistent with the low diffusion rate corresponding to their large size (*Cash* and *Hess* 1981).

2.4 Isolation of Membrane Vesicles that Contain Functional Receptors

A comparison between chemical kinetic measurements with membrane vesicles and electrophysiological measurements with cells can be made,

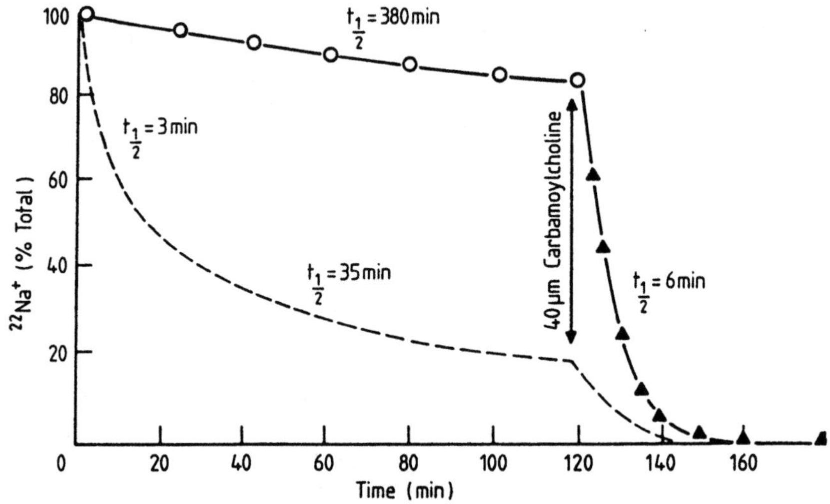

Fig. 8. $^{22}Na^+$ efflux from a heterogenous mixture of vesicles (*dashed line*) and after isolation of vesicles with active receptors (*solid line*), pH 7.0, 4°C. (*Hess* 1979)

providing one can determine the concentration of receptor sites and the internal volume of the vesicles used in the experiments (see Sect. 5) (*Hess* et al. 1981). For this reason, we devised a technique to separate vesicles containing functional receptors from those which do not. The general principle of the method which we developed involved the exchange inside the specific vesicles of a molecule with one density for one outside with a different density. In our case, Cs^+ in the external solution was exchanged for Na^+ inside the vesicles. Vesicles in which this exchange occurs rapidly can be separated, on the basis of their different density, from vesicles in which the exchange occurs slowly. Differential exchange can be obtained either because of differences in passive permeability between vesicles or by taking advantage of the existence of a specific protein which mediates the exchange, in our case the acetylcholine receptor. If the protein responsible for transmembrane translocation is uniformly distributed on the membrane surface, then exchange rate and hence vesicle density are both inversely proportional to the radius of the vesicles. By allowing the exchange to occur for different periods of time, vesicles of different sizes but all containing receptors can be separated from each other. At least three other examples of the general strategy which we adopted to separate vesicles (*Hess* and *Andrews* 1977) have now been published (*Goldin* et al. 1980).

In the separation of vesicles that contain active receptors, we took advantage of the impermeability of these vesicles to inorganic ions in the absence of an activating ligand. We used this kinetic distinction to fill the vesicles without functional receptors with cesium chloride, while the vesicles with functional receptors were filled with a sodium chloride solution (*Hess* and *Andrews* 1977). The sodium-chloride-filled vesicles were then separated from vesicles filled with heavier cesium chloride on the basis of their density in a continuous sucrose-190 m*M* cesium chloride density gradient. The results of such a density gradient experiment are shown in Fig. 7. A comparison of $^{22}Na^{+}$ efflux before and after purification of the vesicles is shown in Fig. 8.

3 Characterization of Vesicles that Contain Functional Receptors

3.1 Does the Purified Vesicle Preparation Contain Vesicles Without Receptors?

Two types of experiments were performed to show that the purified preparation was essentially free from vesicles without functional receptors. The purified vesicles were equilibrated with tracer ions in two ways: (a) incubation for up to 30 s in the presence of 1 m*M* carbamoylcholine to load only those vesicles that contained active receptors and (b) incubation with $^{86}Rb^{+}$ for 20 h to equilibrate all the vesicles. When the release of tracer ions was subsequently measured, it was found that the vesicles which had equilibrated for 20 h with $^{86}Rb^{+}$ contained a somewhat ($\sim$5%) higher tracer ion content than the vesicles equilibrated with $^{86}Rb^{+}$ in the presence of 1 m*M* carbamoylcholine. This difference in tracer ion content was found to be due to a small population of vesicles, which are rather impermeable to inorganic ions and which do not contain functional receptors (*Kim* and *Hess* 1981). In the influx experiments described here, these vesicles do not contribute to the measurement.

3.2 Do the Purified Vesicles Have Similar Kinetic Properties?

A comparison of influx measured using either the heterogeneous vesicle population or the purified vesicles (and the quench flow technique using $^{86}Rb^{+}$ and 1 m*M* carbamoylcholine; see Fig. 7) gave kinetic parameters which agreed within experimental error (*Leprince* 1981; *Sachs* 1982). This indicates that the vesicle population which is used in the measurements contains a narrow distribution of the receptor concentration to internal volume ratio which determines the observed rate of ion flux. The uniformity of the kinetic parameters observed with *E. electricus* vesicles is not observed in vesicles prepared from *Torpedo* spp. electroplax (*Bernhardt* and *Neumann* 1978, 1980; *Walker* et al. 1981; *Hess* et al. 1982).

3.3 Determination of the Internal Volume of Vesicles

At high concentrations of carbamoylcholine or acetylcholine the tracer ions equilibrate across the vesicle membranes within seconds, i.e., influx measured over a wide range of rubidium concentrations indicated that the ion was not bound to the vesicles to a significant extent (*Kim* 1980). Using vidicon flame emission spectroscopy, we showed that the amplitudes of the influx of Na^+, K^+, Li^+, and $^{86}Rb^+$ are the same (*Ramseyer* et al. 1981). On the basis of these measurements and from a knowledge of the specific activity of the $^{86}Rb^+$ in the external solution, the internal volume of the vesicles can be determined from the amplitude of the $^{86}Rb^+$ influx (*Kim* 1980). The internal volume of the vesicles that contain functional receptors was found to be 2.0 ± 0.3 μl/mg membrane protein (*Hess* et al. 1981). This value is based on membrane preparations obtained from ten eels.

3.4 Determination of the Value of R_0, the Moles of Receptor per Liter Vesicle Internal Volume

The concentration of α-bungarotoxin sites (*Lee* 1972) in the most active vesicle fractions, as determined with vesicle preparations from seven eels, was 4.8 ± 0.7 pmol/mg of membrane protein. It had been shown previously that the stoichiometry of α-bungarotoxin-binding sites and receptor ligand-binding sites is 1:1 in electroplax membranes (*Fu* et al. 1977; *Bulger* et al. 1977; *Karin* 1980) and that the binding of two ligand molecules is required to initiate ion translocation (*Cash* and *Hess* 1980). Accordingly, the concentration of receptors is 2.4 pmol/mg membrane vesicle protein. Considering the internal volume of the vesicles, the receptor concentration per liter internal vesicle volume, R_0, is 1.2 μM.

4 Analysis of Quench Flow Kinetic Measurements

The rate of the ion translocation process in vesicles is given by:

$$-\frac{dM}{dt} = \bar{J}\ R_0\ [\overline{AL_2}]_t\ M. \tag{1}$$

M represents the concentration of inorganic ions in moles per liter of internal vesicle volume and can be calculated directly from measurements of tracer ion flux. All the other symbols are defined in Table 1. The following points have been considered in the use of this equation.

Table 1

$J_A = \bar{J}\, R_o\, [\overline{AL_2}]_o$ $(J_A\,(\max) = \bar{J}\, R_o\, (1+\Phi)^{-1})$	J_A is the rate coefficient associated with the active form of the receptor (A)
$J_I = \bar{J}\, R_o [\overline{AL_2}]_\infty$ $(J_I\,(\max) = \bar{J}\, R_o\, K_{c_2}\, \Phi^{-1})$	J_I is the rate coefficient associated with the „inactivated receptor," i.e., the equilibrium mixture of active and inactive receptor forms ([J] + [A])
$[\overline{AL_2}]_o = \dfrac{[L]^2}{[L]^2\,(1+\Phi) + [L]2\ K_1\Phi + K_1{}^2\Phi}$	$[\overline{AL_2}]$ and $[\overline{AL_2}]_\infty$ are the fraction of receptor in the open channel form at t = 0 and at equilibrium respectively
$[\overline{AL_2}]_\infty = \dfrac{L^2\, K_{c_2}}{K_{c_2}\,([L]^2\,(1+\Phi) + [L]\,2\,K_1\,\Phi + K_1{}^2\Phi) + \Phi\,([L]^2 + [L]2\,k_2)}$	
$\Phi = \dfrac{[AL_2]}{[\overline{AL_2}]}$	where AL_2 is the fraction of active form of receptor with two ligand molecules bound and the channel closed
$\alpha = \dfrac{k_{43}\,[L] + k_{21}\,2K_2}{[L] + 2K_2} + \Phi \left[\dfrac{k_{34}[L]^2 + k_{12}[L]\,2\,K_1}{[L]^2\,(1+\Phi) + [L]2\,K_1\Phi + K_1^2\,\Phi}\right]$	
$\alpha\,(\max) = k_{34}\,\Phi\,(1+\Phi)^{-1}$ $\alpha\,(\min) = k_{21}$	
$K_1 = \dfrac{2\,[A]\,[L]}{[AL]} = \dfrac{[AL]\,[L]}{2\,[AL_2]}$	K_1 is the intrinsic dissociation constant of the complexes involving the active (A) form of the receptor, and AL and AL_2 represent species with one and two bound ligand molecules respectively.
$K_2 = \dfrac{[IL]\,[L]}{[IL_2]}$	K_2 is the intrinsic dissociation constant of the complexes involving the inactive (I) form of the receptor and IL and IL_2 represent forms with one and two bound ligand molecules respectively.
$K_{c1} = [AL]\,/\,[IL] = \dfrac{k_{21}}{k_{12}}$	
$K_{c2} = [AL_2]\,/\,[IL_2] = \dfrac{k_{43}}{k_{34}}$	K_{c1} and K_{c2} represent equilibrium constants and lowercase k's the rate coefficients for the interconversion between A and I states.

Most measurements of ion translocation rates have been made using $^{86}Rb^{+}$. We have shown, using vidicon flame emission spectroscopy (*Ramseyer* et al. 1981), that the ion translocation rates of the radioactive tracer ions are the same, within experimental error, as those of Na^{+} and K^{+}.

Experimentally, we measure a rate coefficient, J_A, that is characteristic of the active conformation of the receptor. We have shown that J_A is independent of the concentration of NaCl and KCl in the physiologically important concentration region, 0.024 to 0.36 *M* salts (*Hess* et al. 1981). J_A (max) was found to depend on the activating ligand used, and it is much higher at saturating concentrations of acetylcholine than of carbamoylcholine (*Cash* et al. 1980, 1981). J_A (max) can, however, be expressed as $\bar{J} R_0 (1 + \Phi)^{-1}$, where R_0 represents moles of receptor sites per liter internal vesicle volume and Φ the channel-closing equilibrium constant (see Table 1).

$\bar{J}$ is the specific reaction rate (*Hammes* 1978) of the acetylcholine-receptor-controlled ion translocation. It is a constant that is characteristic of the receptor and is independent of the type of activation ligand used and of other properties of the cells or vesicles that also determine the rates of ion translocation. Its value is expressed in liters per mole per second.

It is customary to express transport of substances across membranes in terms of flux coefficients, which requires a knowledge of the surface area of the vesicles. The internal volume of the vesicle population is obtained from the equilibrium concentration of tracer ions in the vesicles (see Sect. 3), while determination of the surface area requires a knowledge of the average diameter of the vesicles. It is for these reasons that we express our data in terms of rate coefficients. The relations among these rate coefficients, flux coefficients, and the single channel conductance are discussed in Sect. 7.

The importance in Eq. (1) of the term $[\overline{AL_2}]_t$, the fraction of the receptor in the open channel form at time t, is seen from the data in Fig. 13a. One can detect three processes:

1. An initial rapid influx, characterized by a rate coefficient, J_A. This process occurs in the millisecond time region and ends within about 1 s.
2. An influx phase which is slow compared to the initial fast phase, characterized by the rate coefficient, J_I, and which occurs in the second time region. Only a small portion of this phase is shown in the Figure.
3. The conversion of the fast influx into a slow influx, by a process characterized by the rate coefficient α.

The model in Fig. 9c (*Cash* and *Hess* 1980), which relates the ligand-binding process to the transmembrane ion flux, is consistent with the following previous observations.

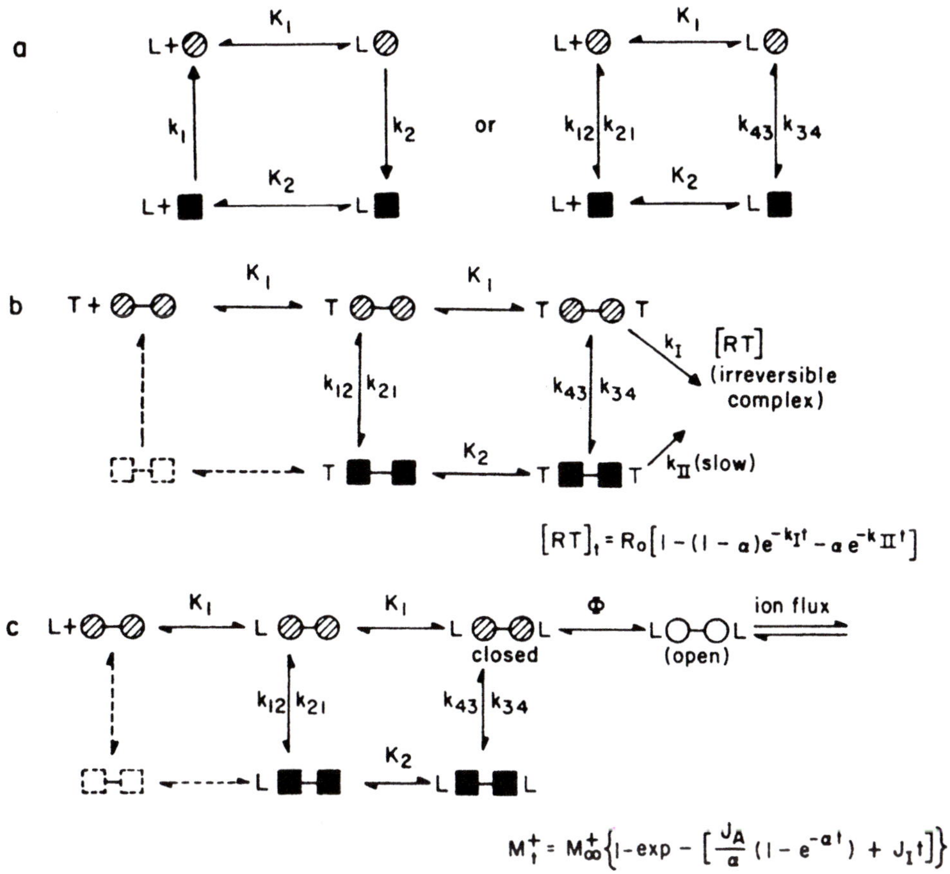

Fig. 9a–c. Minimum mechanisms which have been proposed to account for properties of the acetylcholine receptor. Forms of the receptor directly involved in its physiological function, the active forms, are indicated by ○ and ⊘ and other forms by ■ and ⬚. K and Φ are equilibrium constants, and k is a rate constant for receptor inactivation or of reaction of the receptor with toxin. J_A and J_I are ion flux rate constants associated with the active conformation or equilibrium mixtures of active and inactive forms, respectively, and α is the observed rate constant for receptor inactivation. L represents acetylcholine or carbamoylcholine concentration. The concentrations of the other reactants are represented as follows: T, α-bungarotoxin; R_o, receptor; RT', irreversible receptor-toxin complex; M^+, metal ion concentration inside a cell or membrane vesicle. The subscripts t and ∞ indicate the time at which the measurement was made. The *dotted receptor forms* may exist but are not necessary to account for the measurements. The integrated rate equations which account for the reaction of the receptor over a wide range of toxin concentrations, or for the receptor-controlled ion translocation over a wide range of carbamoylcholine or acetylcholine concentrations, are also shown. The values of the constants pertaining to the various models have been published (*Katz* and *Thesleff* 1957; *Bulger* et al. 1977; *Cash* and *Hess* 1980; *Hess* et al. 1980). **a** Mechanism suggested by *Katz* and *Thesleff* (1957) on the basis of electrophysical measurements with the neuromuscular junction. **b** Minimum mechanism suggested by *Hess* et al. (1975b; *Leprince* et al. 1981) on the basis of kinetic measurements of the reaction of α-bungarotoxin with acetylcholine-receptor-containing membrane vesicles isolated from the electric organ of *E. electricus*. In the equation defining the concentration of the toxin-receptor complex at time t, α is the fraction of sites which bind α-bungarotoxin slowly. **c** Minimum mechanism suggested on the basis of kinetic measurements of the acetylcholine-receptor-controlled ion flux in membrane vesicles isolated from the electric organ of *E. electricus* (*Cash* and *Hess* 1980; *Aoshima* et al. 1980). In the equation defining the concentration of a specific inorganic ion in the vesicles at time t, α is the first-order rate constant for inactivation (desensitization) of the receptor

First, the electrophysiological measurements of *Katz* and *Thesleff* (1957) indicated the interconversion between active and inactive receptor forms and the requirement for the binding of more than one ligand molecule to the receptor for changes in transmembrane voltage to occur (Fig. 9a).

Second, kinetic measurements (*Bulger* et al. 1977) of the specific reaction of α-bungarotoxin a snake toxin (*Lee* 1972), with the receptor in eel membrane vesicles (Fig. 9b) indicated that two toxin molecules are bound to the receptor before the irreversible step in the reaction occurs. The experiments also demonstrated a time-dependent conversion of a low-affinity form of the receptor to a high-affinity form.

Third, measurements of receptor-controlled $^{86}Rb^+$ efflux with electroplax membrane vesicles indicated a rapid efflux phase at high ligand concentration that is followed by a slower phase (Fig. 4) (*Hess* et al. 1978).

Assuming that all the bimolecular steps and the channel opening are fast compared with the protein isomerization steps, which are characterized by the rate coefficients, k_{12}, k_{21}, k_{43}, and k_{34}, the rate equation for the interconversion between active receptor forms (A) and inactive forms (I) is given by:

$$\frac{d\,([A]+[AL]+[AL_2]+[\overline{AL}_2])}{dt} = k_{21}\,[IL] + k_{43}[IL_2] - k_{12}\,[AL] - k_{34}\,[AL_2] \qquad (2a)$$

In our experiments $L_o \gg R_o$, so that we can assume that changes in the ligand concentration are negligible. The relationships between the various species of active (A) and inactive (I) receptor forms and the various constants used in the model are given in Table 1. Using the relationships given in Table 1, substitution in Eq. (2a), and integration gives:

$$\left[[\overline{AL}_2]_t - [\overline{AL}_2]_\infty\right] = \left[[\overline{AL}_2]_o - [\overline{AL}_2]_\infty\right] e^{-\alpha t} \qquad (2b)$$

Substitution of Eq. (2b) into Eq. (1) gives, after integration, the concentration of tracer ions in the vesicles as a function of time:

$$[M]_t/[M]_\infty = 1 - \exp - \bar{J}R_o \left\{\left[[\overline{AL}_2]_o - [\overline{AL}_2]_\infty\right] \left(\frac{1-e^{-\alpha t}}{\alpha}\right) + [\overline{AL}_2]_\infty\, t\right\} \qquad (3)$$

5 Evaluation of the Rate and Equilibrium Constants

5.1 Evaluation of J_I $(= \bar{J}\,R_o\,[\overline{AL}_2]_\infty)$

When the membrane vesicles are incubated with a ligand for longer periods (in practice, 1 h), $e^{-\alpha t}$ in Eq. (3) decays, and the receptor-controlled influx follows a single expontential rate law. J_I can then be evaluated:

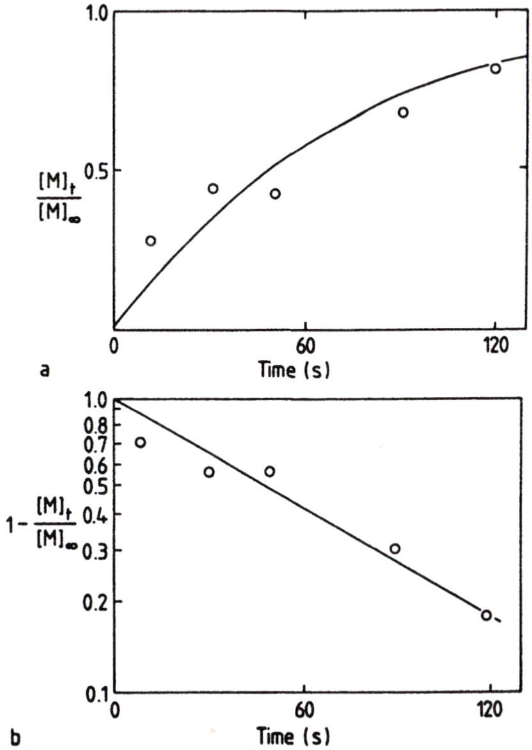

Fig. 10a,b. Ion flux with inactivated (desensitized) receptor. **a** The *E. electricus* vesicles were incubated with ligand for 1 h before influx of $^{86}Rb^+$ was followed at the same ligand concentration as in the incubation. ○, 5 m*M* carbamoylcholine. The line was computed using Eq. (3a) and the semilogarithmic plot shown in **b**. The points were determined using the Millipore assay after dilution into a solution of *d*-tubocurarine. **b** Semilogarithmic plot of the data in **a** according to Eq. (3a). The slope of the line gives the value of J_I as 0.01 s^{-1}

$$[M]_t = [M]_\infty (1 - \exp - J_I t). \tag{3a}$$

Experiments with 5 m*M* carbamoylcholine designed to evaluate J_I are shown in Fig. 10. The $^{86}Rb^+$ influx in a graph of $[M]_t/[M]_\infty$ is plotted as a function of time, where M refers to the $^{86}Rb^+$ content of the vesicles at time t and at the end of the experiment (∞) (Fig. 10a). Figure 10b shows a semilogarithmic plot of the data in Fig. 10a, according to Eq. (3a). A single exponential rate law is obeyed and the slope of the line is directly proportional to J_I. The values of J_I determined over a 2000-fold range of carbamoylcholine concentration are summarized in Fig. 16a. The value of J_I(max) obtained on saturating the receptor with the ligand is obtained from the data shown in this Figure. The relationship between J_I and the ligand concentration is given in Table 1.

5.2 Evaluation of α

When the membrane vesicles are incubated with ligand for short periods of time (10 ms to seconds) and the receptor-controlled influx of tracer ion is measured, α in Eq. (2b) can be evaluated from single influx measurements. In these experiments, vesicles are first mixed with carbamoylcholine and incubated for different periods of time. In the quench flow apparatus, incubation periods as short as 10 ms can be used. Then the vesicles are mixed with tracer ions and carbamoylcholine to determine the amount of ion flux activity remaining. When saturating concentrations of carbamoylcholine are used in this assay procedure and influx is allowed to proceed for a constant period of time, each of the nonexponential terms in Eq. (2b) can be evaluated from a single influx measurement. At saturating concentrations of carbamoylcholine, the influx rate due to the active form of the receptor (characterized by J_A) is larger by a factor of $\sim$500 than the influx rate due to inactive receptor (characterized by J_I). In the assay of receptor activity, an influx period of 1.2 s is used to obtain a maximum influx of tracer ions without having to consider the slow influx phase (which has a $t_{1/2}$ value of about 60 s at saturating carbamoylcholine concentrations). Under these assay conditions, the term containing $(\overline{AL_2})_\infty$ in Eq. (3) can be neglected. Because influx is measured for a constant period of time, $(1 - e^{-\alpha t})/\alpha$ and $\bar{J}\ R_0$ can be represented by a constant, K. Equation (3) reduces to the following:

$$[\overline{AL_2}]^T = -\ln \frac{[M]_\infty - [M]_{t\,=\,\mathrm{constant}}^{\,T}}{[M]_\infty} \quad K \tag{3b}$$

where $[M]_\infty$ has the usual meaning and $[M_{t\,=\,\mathrm{constant}}]^T$ is the concentration of tracer ions after a constant influx period at saturating concentrations of carbamoylcholine. The superscript T represents the preincubation time with a particular concentration of carbamoylcholine. All the terms in Eq. (2b) can be evaluated by use of Eq. (3b). For instance, the term $[\overline{AL_2}]_0$ is obtained from 1.2-s influx measurements performed without prior preincubation of the receptor with carbamoylcholine.

The decrease in receptor activity after various periods of preincubation with carbamoylcholine is shown in Fig. 11. The ordinate of the graph gives the $^{86}Rb^+$ content of the vesicles after an influx period of 1.2 s at a saturating concentration of ligand (10 m*M* carbamoylcholine) after various periods of preincubation $[[M]_{t\,=\,1.2\,s}^T]$. The incubation periods are given on the abscissa of the graph. The $[M]_{t\,=\,1.2\,s}^T$ values were normalized by the $^{86}Rb^+$ content of the vesicles obtained during influx for 1.2 s without any preincubation $[[M]_{t\,=\,1.2\,s}^{T\,=\,0}]$. The carbamoylcholine concentrations during preincubation were 10 m*M* (□), 2 m*M* (○), and 0.25 m*M* (△).

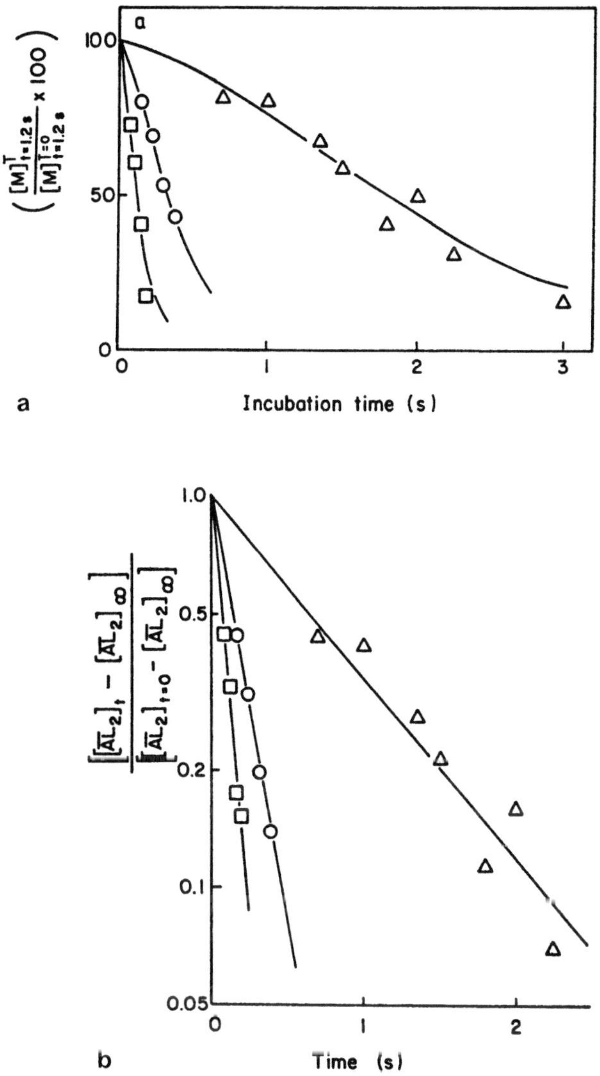

Fig. 11a,b. The effect of carbamoylcholine concentration on the rates of receptor inactivation. $M^T_{t=1.2\ s}$ refers to the concentration of $^{86}Rb^+$ inside the vesicles 1.2 s after influx was initiated by addition of $^{86}Rb^+$ and carbamoylcholine. The superscript refers to the duration of the first incubation with carbamoylcholine. **a** Inhibition of receptor-controlled flux by incubation with carbamoylcholine for various periods of time: △, 0.25 m*M*; ○, 2 m*M*; □, 10 m*M*. Each point is the mean of three determinations. The solid lines were calculated using Eq. (3c) and the values of α determined in the experiments illustrated in **b**. **b** Evaluation of the rate constant for the inactivation reaction. The data shown in a were replotted according to Eq. (3c). The values of α obtained from the slope of the lines at the different concentrations of carbamoylcholine are: △, at 0.25 m*M* carbamoylcholine, 0.9 ± 0.1 s^{-1}; ○, at 2 m*M* carbamoylcholine, 4.9 ± 0.1 s^{-1}; □, at 10 m*M* carbamoylcholine, 8.8 ± 0.4 s^{-1}. (*Aoshima* et al. 1981)

Figure 11b is a semilogarithmic plot of the data according to Eq. (3c). This equation is obtained by substituting Eq. (3b) in Eq. (2b). A single exponential rate law is obtained and the slope

$$\ln \frac{[M]_\infty - [M]^{T}_{t=\text{constant}}}{[M]_\infty - [M]^{T=0}_{t=\text{constant}}} = e^{-\alpha t} \tag{3c}$$

of the line (Fig. 4b) is directly proportional to α. The values of α determined over a 2000-fold range of carbamoylcholine concentration are summarized in Fig. 16b.

The value of α(max), obtained on saturating the receptor with ligand, is obtained from the data shown in the Figure. The relationship between α and ligand concentration is given in Table 1.

It is possible to obtain the values of k_{21} (see Fig. 12) from reactivation experiments. In these measurements, the receptor is incubated with low concentrations of ligand until an equilibrium between active and inactive receptor forms is obtained. Subsequently, the carbamoylcholine concentration is reduced sevenfold by diluting the incubation mixture. After various periods of time in the dilution buffer, vesicles are mixed with tracer ions and influx is measured for a constant period of time (1.2 s) at saturating concentrations of carbamoylcholine. In the experiment in Fig. 12a, vesicles incubated with 70 μM carbamoylcholine for 1 h were diluted to 10 μM carbamoylcholine. After various periods of time, $^{86}Rb^+$ influx was allowed to proceed in the presence of 5 mM carbamoylcholine for 1.2 s. In this time interval, only the fast influx phase due to the active state of the receptor was observed. The ordinate of the graph gives the concentration of $^{86}Rb^+$ in the vesicles in terms of counts per minute (cpm). The solid circles represent the influx measured after the vesicles had been incubated in the dilution buffer for the periods of time given on the abscissa of the graph. The open triangle represents a control experiment in which the vesicles were not exposed to carbamoylcholine. The open circle represents an experiment in which the concentration of carbamoylcholine in the preincubation solution and in the dilution buffer was the same (70 μM carbamoylcholine). The difference in the $^{86}Rb^+$ content of the vesicles in the two experiments designated by △ and ○ reflects the equilibrium between active and inactive receptor states. In 70 μM carbamoylcholine, about 30% of the receptor remains in the active state, in agreement with the results obtained previously (*Aoshima* et al. 1981). The solid triangles represent another control experiment in which preincubation with 70 μM carbamoylcholine was omitted. The difference between the tracer ion content of the vesicles after 20 s in the dilution buffer and the experiment designated by ▲ reflects the fraction of receptors in the inactive state, about 15% in the dilution buffer which contains 10 μM ligand.

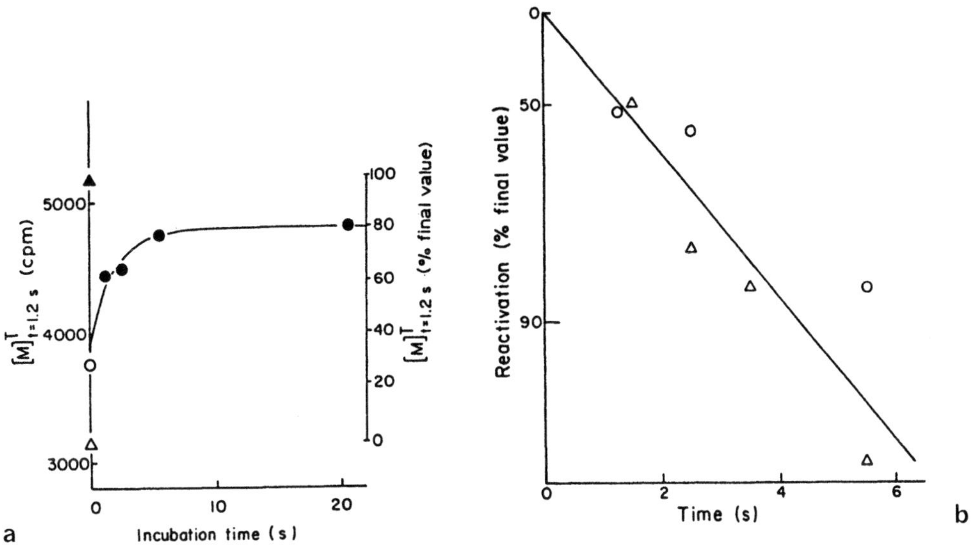

Fig. 12a,b. a Reactivation of the inactivated receptor. The vesicles were exposed to 70 μM carbamoylcholine for 1 h. The vesicles were then diluted to give a final carbamoylcholine concentration of 10 μM. After the vesicles were exposed for various periods of time to the dilution buffer, influx was allowed to proceed for 1.2 s in the presence of 5 m*M* carbamoylcholine. $M_{t=1.2\ s}$ on the ordinate refers to the concentration of $^{86}Rb^+$ in the vesicles (in cpm) after the various periods of time in the dilution buffer given on the abscissa of the graph. ●, vesicles in the dilution buffer (10 μM carbamoylcholine) for the times indicated on the abscissa. The solid line was computed using an α value of 0.46 s^{-1}. Control experiments: ○, dilution buffer with the same concentration of carbamoylcholine as the preincubation solution (70 μM). ▲, vesicles not exposed to carbamoylcholine before $^{86}Rb^+$ influx was measured. △, passive $^{86}Rb^+$ influx, i.e., in the absence of carbamoylcholine. **b** Evaluation of rate constants of the reactivation reaction. The data from two experiments (△, ○) were plotted according to the following equation:

$$\frac{[\overline{AL}_2]_\infty - [\overline{AL}_2]_t}{[\overline{AL}_2]_\infty - [\overline{AL}]_{t=0}} = e^{-\alpha t}$$

The *abscissa* gives the time the vesicles remain in the dilution buffer, and the *ordinate*, percent reactivation corresponds to the left-hand term of the equation multiplied by 100. The value obtained from the slope of the line is 0.46 ± 0.05 s^{-1}. (*Aoshima* et al. 1981)

The data in Fig. 12a can be used to evaluate α at low concentrations of carbamoylcholine and to obtain an estimate of the isomerization rate constant, k_{21} (see Table 1). Figure 12b shows a semilogarithmic plot of the data, according to Eq. (2b), rearranged for reactivation of the receptor. A single exponential rate law is obtained and the slope of the line (Fig. 12b) is directly proportional to α. A value for α of 0.46 s^{-1} is obtained from the experiment shown.

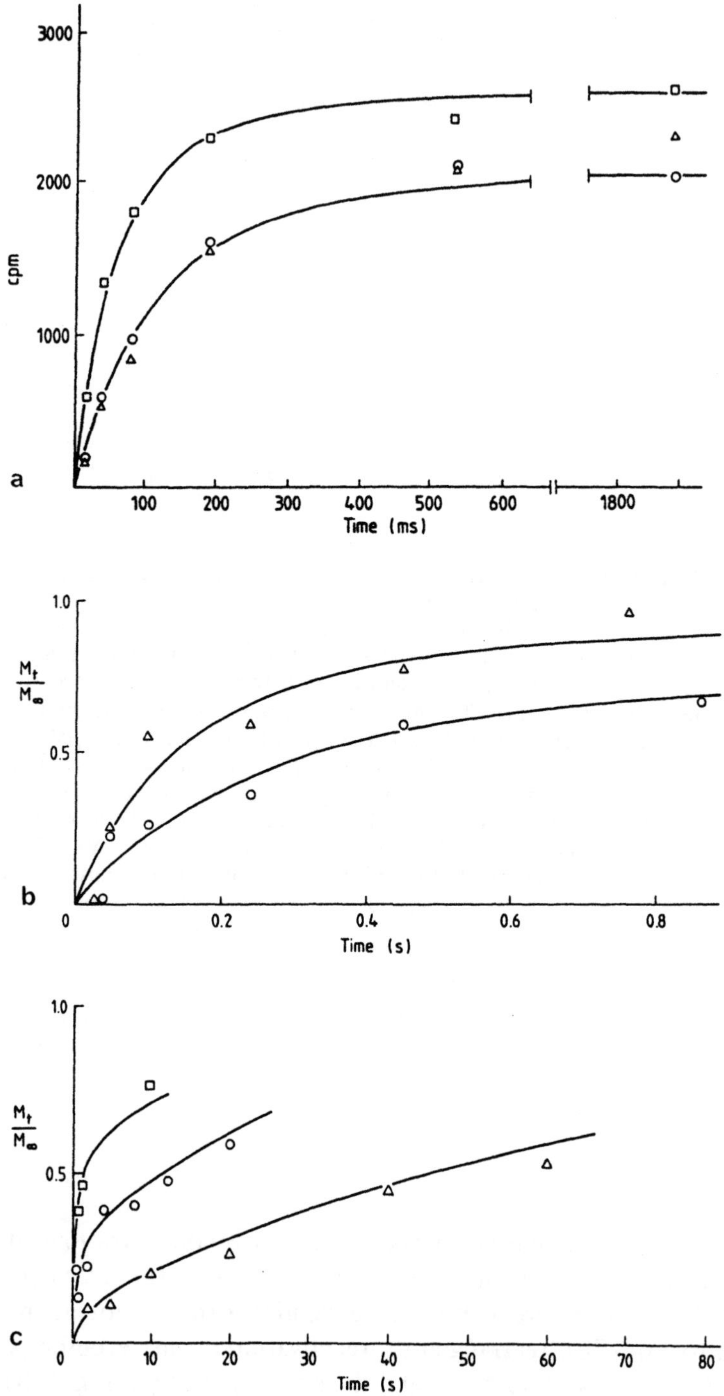
3000
2000
1000
cpm
a
100
200
300
400
500
600
1800
Time (ms)
1.0
0.5
$\frac{M_t}{M_\infty}$
b
0
0.2
0.4
0.6
0.8
Time (s)
1.0
0.5
$\frac{M_t}{M_\infty}$
c
0
10
20
30
40
50
60
70
80
Time (s)

Fig. 13a–c

5.3 Evaluation of J_A [$= \bar{J}\, R_o\, [\overline{AL_2}]_o$]

The values of J_A, characteristic of the active form of the receptor, can be obtained from influx measurements (shown in Fig. 13). With the constants J_I and α determined independently, J_A is the only unknown term in the integrated rate equation [Eq. (3)]:

$$[M]_t/[M]_\infty = 1 - \exp - \bar{J} R_o \left\{ \left[[\overline{AL_2}]_o - [\overline{AL_2}]_\infty \right] \left(\frac{1 - e^{-\alpha t}}{\alpha} \right) + [\overline{AL_2}]_\infty t \right\} \quad (3)$$

which was used to calculate the solid curves in Fig. 13 and to evaluate J_A.

Evaluation of J_A depends on an independent evaluation of α or a knowledge of M_∞, the tracer ion content of the vesicles after equilibration with tracer ions in the external solution. The value of M_∞ cannot always be determined easily. Inspection of the results obtained with saturating concentrations of carbamoylcholine (10 m*M* and 30 m*M*) in Fig. 13a illustrates this point. The results of experiments in which acetylcholine (upper curve of Fig. 13a) was the activating ligand make it clear that the equilibrium concentration of tracer ions (M_∞) was not reached during the period of observation in the experiments in which carbamoylcholine was used. The slow phase of the receptor-controlled influx at saturating ligand concentration characterized by J_I ($t_{1/2}$ ~1 min), and the nonspecific slow exchange of ions ($t_{1/2}$ ~3 min) complicate the measurements (*Kim* and *Hess* 1981). This is so particularly when the slow receptor controlled

---

Fig. 13a–c. Acetylcholine-receptor-mediated ion flux ($^{86}Rb^+$ tracer) into the membrane vesicles. **a** Saturating concentrations of activating ligand were used: □, 0.8 m*M* acetylcholine; △, 10 m*M* carbamoylcholine; and ○ 30 m*M* carbamoylcholine in outside *E. electricus* Ringer's solution. The background (1300 cpm) was constant within this time range. The flow rate of the solutions before mixing in the quench flow apparatus was 1.1 ml/s; pulsed mode was used above and continuous mode below 400 ms reaction time (*Cash* and *Hess* 1981). The lines were computed using Eq. (3).

Upper line: $J_A = 16\ s^{-1}$; $\alpha = 5\ s^{-1}$
Lower line: $J_A = 7\ s^{-1}$; $\alpha = 5\ s^{-1}$
$M_\infty = 2700$ cpm.

The process characterized by J_I occurs on a much slower time scale and only a small part of this phase is shown.

b, c Lower carbamoylcholine concentrations. Each point is the mean of three determinations. M_t/M_∞ is the fraction of the total ion exchange that has occurred in time t. Ligand concentration in **b**: △, 2 m*M*; ○, 1 m*M*; in **c**: □, 0.5 m*M*; ○, 0.25 m*M*; △, 0.1 m*M*. The lines were computed using Eq. (3) and the values of the constants given in Fig. 15. The values α and J_A obtained at different concentrations of carbamoylcholine are summarized in Fig. 16b,c. (*Cash* and *Hess* 1980) influx is only a small fraction of the total influx [corresponding to the difference between the two curves which show tracer ion influx during the first 0.5 s (Fig. 13a)]. Underestimation of M_∞ leads to an overestimation of J_A and an underestimation of α. This problem was encountered in our first measurements of carbamoylcholine-induced influx (*Cash* and *Hess* 1980).

An additional complication in evaluation of J_A occurs at low concentrations of ligand. Equation (3) shows that the amplitude of the influx due to the phase characteristic of the active form of the receptor depends on J_A/α. As can be seen in Fig. 16b,c, α and J_A have different dependencies on carbamoylcholine concentration; α is larger than J_A at low ligand concentrations and smaller than J_A at high ligand concentrations. This means that at intermediate ligand concentrations, influx measurements reflect the value of J_I and J_A. An example of this is given in Fig. 13c, where influx of $^{86}Rb^+$ in the presence of 100 μM carbamoylcholine is shown (△). At this concentration, the calculated amplitude of the influx ascribed to the active state of the receptor is about 5% of the total amplitude and the value of J_A is 0.15 s^{-1}, corresponding to a $t_{1/2}$ for influx of 6 s. The apparent $t_{1/2}$ value in the experiment shown in Fig. 13c (△) appears to be 60 s, characteristic of the value of J_I. Direct measurements of J_I at a carbamoylcholine concentration of 100 μM (Fig. 10a) give a $t_{1/2}$ value of 86 s. In the absence of independent measurements, the inactivation process may be missed under the experimental conditions and wrong conclusions will be reached, namely that J_A is measured and that the receptor is not inactivated at low ligand concentrations.

These problems can be avoided by determining α and J_I independently, using conditions in which each of these processes follows single exponential rate laws. With the values of α and J_I determined independently, J_A is the only parameter remaining to be determined (Fig. 13). A nonlinear computer program was used in the evaluation of J_A to obtain an optimal fit of the data to theory (*Bevington* 1969).

5.4 Evaluation of K_1, Φ, and $\bar{J}\,R_o$

The relationship among J_A [$= \bar{J}\,R_o\,(\overline{AL_2})_o$], the constants K_1 and Φ, and ligand concentration is given by

$$J_A = \bar{J}\,R_o \frac{L^2}{L^2\,(1+\Phi)+2\,L\,K_1\,\Phi+K_1^2\,\Phi} \tag{4a}$$

and allows evaluation of $\bar{J}$, K_1, and Φ. Equation (4a) can be written in a linear form (*Hess* et al. 1981):

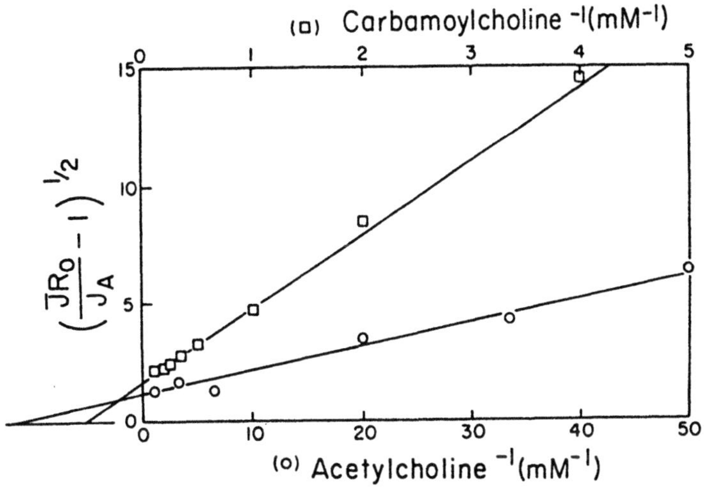

Fig. 14. Acetylcholine-receptor-mediated ion influx. A plot of the J_A values obtained with carbamoylcholine (□) or acetylcholine (○) as the activating ligand, according to Eq. (4b). This plot indicates a square dependence of transmembrane ion flux on ligand concentration and can be used to determine the values of $\bar{J} R_0$, K_1, and Φ. The slope gives $K_1 \Phi^{1/2}$, the ordinate intercept gives $\Phi^{1/2}$, and the abscissa intercept gives $1/K_1$. For carbamoylcholine, K_1 = 1.9 m*M*, Φ = 2.8. For acetylcholine, K_1 = 80 *μM*, Φ = 1.5. The coordinates of the lines and the value of $\bar{J} R_0$ were obtained by use of a least squares computer program. In the case of both acetylcholine and carbamoylcholine, an optimal fit of the data to Eq. (4b) was obtained with a $\bar{J} R_0$ value of 37 s^{-1}, χ^1 = 0.8. (*Bevington* 1969)

$$\left(\frac{\bar{J} R_0}{J_A} - 1\right)^{1/2} = \Phi^{1/2} + \Phi^{1/2} K_1 [L]^{-1} \tag{4b}$$

The J_A values obtained with either carbamoylcholine or acetylcholine as the activating ligand are plotted according to Eq. (4b) in Fig. 14. This equation is derived on the basis of two ligand binding sites characterized by the microscopic dissociation constant K_1. The values of K_1 and Φ are obtained from the abscissa and ordinate intercepts of the plot, respectively. In the case of both acetylcholine and carbamoylcholine, an optimal fit of the data to Eq. (4b) is obtained with a $\bar{J} R_0$ value of 37 s^{-1}.

Separation of vesicles that contain functional receptors from those that do not permits the determination of the receptor site concentration in the vesicle population with which the measurements were made. We were then able to determine that the value of R_0 is 1.2 *μM* (see Sect. 2). Accordingly, $\bar{J} = 3 \times 10^7\ M^{-1}\ s^{-1}$ at pH 7.0 and 1°C.

5.5 Evaluation of K_2 and K_{c2}

The relationship between $J_I = [\bar{J}\, R_o\, (AL_2)_\infty]$ and the constants K_2 and K_{c2} is given by:

$$J_I = \bar{J}\, R_o \frac{L^2 K_{c2}}{K_{c2}\, (L^2\,(1+\Phi) + L\, 2\, K_1\, \Phi + K_1^2\, \Phi) + \Phi\, (L^2 + L\, 2\, K_2)}. \quad (5a)$$

A combination of Eqs. (4a) and (5a) can be written in linear form (5b):

$$\left(\frac{\bar{J}\, R_o}{J_I} - \frac{\bar{J}\, R_o}{J_A}\right) = \frac{\Phi}{K_{c2}} + \frac{\Phi\, 2\, K_2}{K_{c2}}\, \frac{1}{L}. \quad (5b)$$

A plot of $1/L$ versus the left-hand side of Eq. (5b) gives ordinate and abscissa intercepts from which the values of K_2 and Φ/K_{c2} can be estimated.

5.6 Evaluation of K_{c1}, k_{43}, and k_{12}

The value of K_{c1} is determined by the values of K_1, K_2, and K_{c2} ($K_{c1} = K_1\, K_{c2}/K_2$). Since we can estimate the values of k_{34} and k_{21} from determinations of α at saturating ligand concentrations (Table 1) and at low ligand concentration, respectively (see Sect. 5.2), we can obtain estimates for the rate constants k_{43} and k_{12}, as well.

6 Minimal Mechanism for *Electrophorus electricus* Acetylcholine-Receptor-Controlled Ion Flux

We have proposed (*Cash* and *Hess* 1980; *Aoshima* et al. 1980) a minimal kinetic scheme (Fig. 15) for acetylcholine-receptor-mediated ion flux in vesicles prepared from *E. electricus* electroplax membranes. This scheme accounts for the measurements made of ion flux without prior inactivation (desensitization) of the receptor, ion flux after inactivation, inactivation of the receptor, and reactivation of the inactivated receptor, over a 2000-fold range of carbamoylcholine concentration. The receptor can exist in two states, active (A) and inactive (I), both of which bind activating ligands. We refer to the equilibrium mixture of active and inactive receptor forms at any given ligand concentration as the inactivated receptor. The ligand-binding and channel-opening processes are fast, relative to the ion flux equilibration, and appear as equilibria. The interconversions between the active and inactive states are first order, with rates comparable with the ion flux equilibration and may be measured by this process.

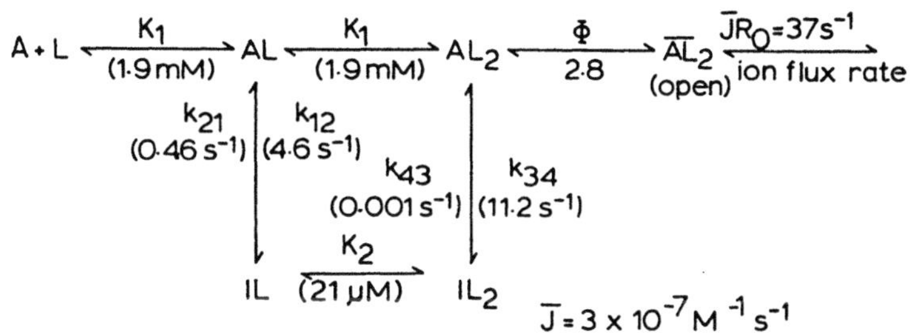

$$[M]_t/M_\infty = 1 - \exp -\bar{J}R_0\left[(\overline{AL}_2)_0 \frac{1-e^{-\alpha t}}{\alpha} + (\overline{AL}_2)_\infty t\right]$$

Fig. 15. The minimum model which relates ligand-binding steps, channel-opening steps, and rates of interconversion between active (*A*) and inactive (*I*) receptor states, to rates of receptor-controlled ion flux. The constants were determined at pH 7.0 at 1°C, with an outside eel Ringer's solution both inside and outside the membranes and with carbamoylcholine as the activating ligand

Channel opening occurs as a rearrangement when two ligand molecules are bound to the active state. This allows ion flux through the channel, characterized by $\bar{J}$ R_0, which is a process of a different kind, not involving depletion of receptor species (A, AL, etc.). Channel opening with one bound ligand molecule is negligible. In contrast, inactivation with one as well as with two bound ligand molecules is significant. Thus the scheme accounts for the different ligand dependencies of ion flux and receptor inactivation. The cyclic features of the scheme allow the reactivation rate of the receptor to be independent of the inactivation process, as was pointed out by *Katz* (*Katz* and *Thesleff* 1957). The channel-opening process, characterized by the equilibrium constant Φ, perturbs the equilibrium among species A, AL, and AL_2 and thereby affects the binding of ligand to the active state of the receptor. The absence of free I in the scheme implies that its contribution is not significant, rather than that it cannot be formed. The absence of inactive receptor in absence of bound ligand is not excluded on the basis of structural considerations. The presence of $\overline{IL}_2$, an open-channel form of the inactivated state, is not required to account for the measurements available at present.

6.1 Consistency with Chemical Kinetic Measurements

The model differs from all other models previously proposed (for recent reviews see *Adams* 1981; *Changeux* 1981), in that all the equilibrium constants and protein isomerization rate constants pertaining to the interconversion between active and inactive receptor states can be evaluated, and that the model and the constants account for the ion translocation process over a wide range of ligand concentrations used. The constants are given in

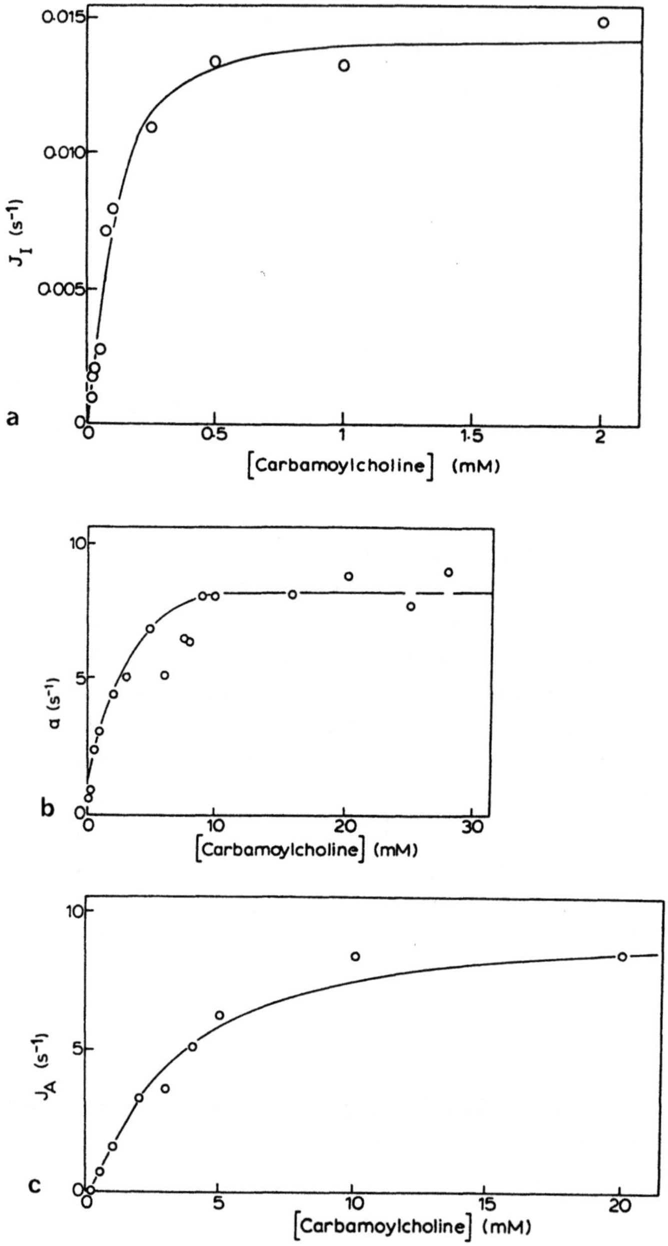

Fig. 16a–c. The dependencies on carbamoylcholine concentration of the responses of the acetylcholine receptor at pH 7.0 and 1°C. The first order rate coefficients of **a** influx of $^{86}Rb^+$ after inactivation (desensitization) mediated by the equilibrium mixture of *active* and *inactive* states (*Aoshima* et al. 1981); **b** inactivation (desensitization) of the receptor (*Aoshima* et al. 1981), and **c** influx of $^{86}Rb^+$ with receptor in the active state (*Cash* and *Hess* 1980).

The constants listed in Fig. 15 and the integrated rate equation (3) pertaining to the model (Fig. 15) were used to calculate the *solid lines*

Fig. 15 for carbamoylcholine. These constants adequately account for the concentration dependence of the four types of measurements made. The solid curves in Fig. 16a–c represent the calculated values for (a) ion flux mediated by inactivated (desensitized) receptor (the equilibrated active and inactive states) (Fig. 16a), (b) rates of equilibration of active and inactive receptor forms (Fig. 16b), and ion flux mediated by the active receptor before inactivation occurs (Fig. 16c). As can be seen from the figure, the calculated values of J_I, α, and J_A account for all the measurements we have made over a 2000-fold range of carbamoylcholine concentration. The same agreement was obtained in measurements with acetylcholine over a 5000-fold concentration range (*Cash* et al. 1981).

6.2 Ligand-Binding Measurements

The number of receptor sites in *E. electricus* vesicle preparations is several hundred times lower than that in *Torpedo* spp. preparations and therefore the preparation is not suitable for binding measurements with ligands that bind reversibly. Direct measurements, in the subsecond time region, of ligand binding to *Torpedo* membrane preparations have been attempted to account for the isomerization of the receptor as a result of ligand binding (for a recent review see *Changeux* 1981). Of particular interest are fluorescent probes that bind reversibly (*Grünhagen* and *Changeux* 1976) and fluorescent analogs of acetylcholine (*Heidmann* and *Changeux* 1978; *Jürss* et al. 1979) because of the possibility of using the same ligand and the same membrane preparation for investigating the mechanism of ligand binding and of ligand-induced receptor-controlled flux. A preliminary model based on ligand binding measurements includes two inactivated states of the receptor, one inactivation process being fast and the other slow (*Grunhagen* et al. 1977; *Heidmann* and *Changeux* 1980). Determination of all the rate and equilibrium constants in this model and comparison with ion translocation measurements using the same vesicle preparation and ligand should be of great interest.

7 Single-Channel Recording Measurements with Muscle Cells and the Theoretical Basis for Comparison with Chemical Kinetic Measurements of Ion Translocation Rates with Membrane Vesicles

Statistical methods for investigations of dynamic properties of acetylcholine-receptor channels in muscle cells were introduced by *Katz* and *Miledi* (1970, 1972, 1977) and refined by *Anderson* and *Stevens* (1973).

An elegant method for making the measurements, the patch clamp technique, has recently been developed by *Neher* and *Sakmann* (1976).

A comparison between chemical kinetic measurements of ion translocation rates and patch clamp recordings of single cells is possible (*Hess* et al. 1984).

A range of carbamoylcholine (2000-fold), acetylcholine (5000-fold), and suberyldicholine (60- to 10 000-fold) concentrations was used to evaluate the specific reaction rate, $\bar{J}$, for the receptor-controlled ion translocation process in *E. electricus* vesicles (*Hess* et al. 1981, 1982). In eel Ringer's solution, at pH 7.0 and 1°C, a value for $\bar{J}$ of 3×10^7 M^{-1} s^{-1} was obtained (see Sect. 5 and Fig. 14).

The single-channel current recording technique (*Neher* and *Sakmann* 1976) was used to determine the single-channel conductance, γ, of the acetylcholine receptor in the *E. electricus* electroplax cells from which the vesicles were prepared. In eel Ringer's solution, at pH 7.0 and room temperature, a value of 53 ps was obtained (*Hess* et al. 1984). This corresponds to a $\bar{J}$ value of 5×10^7 M^{-1} s^{-1}.

8 Ion Translocation in Vesicles Prepared from *Torpedo* Species Electroplax

Acetylcholine-receptor-controlled ion translocation has been studied in vesicles prepared from the electroplax of *Torpedo* spp. (*Hazelbauer* and *Changeux* 1974; *Popot* et al. 1974; *Sugiyama* et al. 1975; *Popot* et al. 1976; *Miller* et al. 1978; *Neubig* and *Cohen* 1980; *Delgeane* and *McNamee* 1980). Typical *Torpedo* spp. vesicles have about 1.5 moles of toxin sites per mg membrane protein (*Karlin* 1980), which exceeds, by several orders of magnitude, the concentration of such sites in *E. electricus* vesicles. At high carbamoylcholine concentrations the observed influx rates exceed the time resolution of the quench flow and stopped flow methods (*Neubig* and *Cohen* 1980), which have been used to investigate the ion translocation process. Reversibly (*Moore* and *Raftery* 1980) and irreversibly (*Neubig* and *Cohen* 1980) binding toxin inhibitors of the receptor have been used to reduce the observed flux rates.

Measurements of the influx of Tl^+ in the presence and absence of histrionicotoxin (*Moore* and *Raftery* 1980), and efflux of $^{22}Na^+$ in the presence and absence of α-bungarotoxin (an irreversible inhibitor), gave results which differed in two important aspects from each other.

First, the effect of carbamoylcholine concentration on the observed flux rate coefficient differed. Measurements of Tl^+ influx in the presence and absence of histrionicotoxin as a function of carbamoylcholine concentration could not differentiate between two alternatives, the requirement for binding of one or two ligand molecules to induce ion flux (*Moore* and *Raftery* 1980). Measurements of $^{22}Na^+$ efflux in the presence and absence of α-bungarotoxin gave a flux rate coefficient vs carbamoylcholine concentration curve which was sigmoidal and gave a Hill coefficient close to 2 (*Neubig* and *Cohen* 1980). The Tl^+ flux measurements are restricted to only high influx rates, since in absence of an activating ligand the permeability of the vesicles to Tl^+ [$t_{1/2}$ a few s (*Moore* and *Raftery* 1980)] is much larger than that to $^{22}Na^+$ ($t_{1/2}$ ∼ 1 h).

Second, in $^{22}Na^+$ efflux measurements, the maximum observed flux rate of 310 s^{-1} (*Hess* et al. 1982b) is about three times lower than in the Tl^+ influx measurements (*Moore* and *Raftery* 1980). However, calculations of the number of ions translocated per receptor site per ms are lower by a factor of 20 than those calculated on the basis of Tl^+ measurements. The former value was based on a determination of toxin sites of the membrane preparation with which the experiments were done, the latter on a number of assumptions. The separation of *Torpedo* spp. vesicles of uniform size and containing functional receptors from the rest of the membrane preparation (which also has toxin sites) has not yet been accomplished. Therefore, calculations of ions translocated per toxin site are expected to be low, as found by *Cohen* and colleagues (*Neubig* et al. 1982). A direct comparison of flux rates obtained with different preparations on the basis of toxin sites is also not possible. Eight *T. californica* preparations (*Hess* et al. 1982) were found to vary by a factor of 15 in the J_A values, while the receptor concentration, as determined with α-bungarotoxin, differed by a factor of only 3. In the measurements shown, we used preparations that were characterized by the same J_A value in the presence of 1 m*M* carbamoylcholine.

A technique has now been developed which avoids the use of extraneous toxins in measuring receptor controlled influx rates that are too high to be measured with the usual quench flow method (*Hess* et al. 1982). Advantage is taken of the natural inactivation of the receptor by its own ligand and of the double mixing technique used to evaluate α. When $J_A \gg \alpha$, influx approximates a single exponential rate law [Eq. (3)] and the value of $(J_A)_{T=0}$ and of α can be evaluated from the values of $(J_A)_T$. $(J_A)_T$ is obtained by first preincubating the vesicles for various periods of time (T) with various concentrations of carbamoylcholine and then measuring $^{86}Rb^+$ influx for 10 ms in the presence of 10 m*M* carbamoylcholine. Under these circumstances, the relationship between $(J_A)_T$, $(J_A)_{T=0}$, and α is given by:

$$(J_A)_T = (J_A)_{T=0} \quad e^{-\alpha t} \tag{6}$$

A semilogarithmic plot of $(J_A)_T$, obtained after various periods of preincubation of the *T. californica* vesicles with carbamoylcholine, vs preincubation time is shown in Fig. 17a. In the presence of 10 m*M* carbamoylcholine, the inactivation rate constant, α, obtained from the slope of the line is $2.0 \pm 0.2\ s^{-1}$. The values of α at lower concentrations of carbamoylcholine, 3 and 0.5 m*M*, are $0.5\ s^{-1}$ and $0.14\ s^{-1}$, respectively. All lines, corresponding to different concentrations of carbamoylcholine during incubation, have a common ordinate intercept, because influx was always measured in the presence of 10 m*M* carbamoylcholine. The intercept gives the value of $(J_A)_{T=0}$ as $310\ s^{-1}$ in the absence of inactivation of the receptor.

Complete curves for $^{86}Rb^+$ influx in 10 m*M* carbamoylcholine obtained after the vesicles were incubated in 10 m*M* carbamoylcholine for 2 s or longer are given in Fig. 17b. The flux rate continues to decrease with the duration of the preincubation, but at about only 6% of the rate at which flux decreased during the first 2 s of preincubation. A considerable amount of flux activity, characterized by the rate coefficient which we designate J_I, is therefore still associated with the acetylcholine receptor after the first inactivation process has gone to completion (Fig. 17b). The value of the rate coefficient J_I is plotted on a logarithmic scale as a function of incubation time in Fig. 17c.

$$(J_I)_T = (J_I)_{T=0}\ e^{-\beta t} \tag{7}$$

The slower inactivation process, characterized by the rate coefficient, β, also appears to follow a first-order rate law. The value ($0.12\ s^{-1}$) of β, obtained from the slope of the line, is about 1/15th the value of α at the same concentration (10 m*M*) of carbamoylcholine. The ordinate intercept gives the value of $(J_I)_{T=0}$ as $1.3 \pm 0.2\ s^{-1}$, which is 300 times smaller than the value of $(J_A)_{T=0}$.

Fig. 17a–c. a Rapid inactivation of flux of $^{86}Rb^+$ into *T. californica* vesicles by different concentrations of carbamoylcholine, pH 7.0, 1°C. The subscript T refers to the length of incubation with carbamoylcholine before influx was measured. The values of $(J_A)_T$ were calculated from 10 ms influx measurements in the presence of 10 m*M* carbamoylcholine and $^{86}Rb^+$ and are plotted on a semilogarithmic scale vs time, according to Eq. (6). Different symbols represent different membrane preparations. Values for α at different concentrations of carbamoylcholine are:

$2 \pm 0.2\ s^{-1}$ (10 m*M* [●, ⊕, ⊡]);
$0.5 \pm 0.04\ s^{-1}$ (3 m*M* [■]);
$0.14 \pm 0.02\ s^{-1}$ (0.5 m*M* [△]).

The mean value of $(J_A)_{T=0}$ obtained from the ordinate intercept of the solid lines is $310 \pm 66\ s^{-1}$. The solid lines were computed by means of Eq. (6) using the value of $(J_A)_{T=0}$ and the values of α obtained with different concentrations of carbamoylcholine in the incubation. Values of χ^2 (*Bevington* 1969) were less than 1.0 in all determinations.

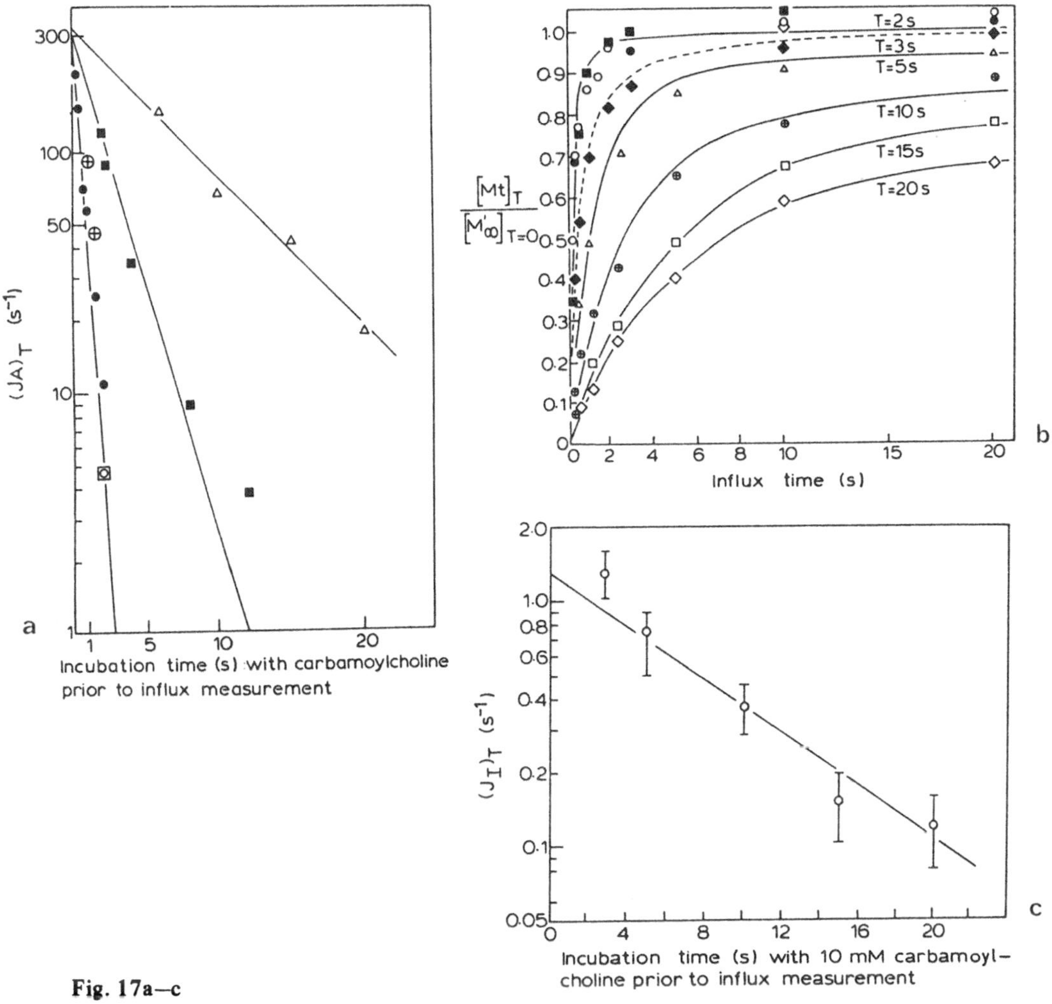

Fig. 17a–c

b, c The second, slower phase of inactivation of the *T. californica* acetylcholine receptor. Acetylcholine-receptor-mediated flux of $^{86}Rb^+$ into vesicles in the presence of 10 m*M* carbamoylcholine, pH 7.0, 1°C, after various relatively long periods of incubation with 10 m*M* carbamoylcholine. M_t/M_∞ is the fraction of influx that has occurred at the time indicated on the abscissa. T refers to the preincubation time which is given on the right-hand side of each of the curves shown.

b Influx after 2 s (○, ●, ▽), 3 s (♦), 5 s (△), 10 s (⊕), 15 s (□) and 20 s (◇) of incubation. Coordinates of influx curves after 2 s incubation were computed by using the values $J_A = 4.5\ s^{-1}$, $\alpha = 2.0$; $\chi^2 = 0.7$. Coordinates of the other curves were computed by use of the second inactivation rate coefficient, β, ($0.12\ s^{-1}$) and the influx rate coefficients given in **c** as a function of incubation time.

c Evaluation of the second inactivation rate coefficient, β, and the influx rate coefficient, $(J_I)_{T=0}$. $(J_I)_T$ values, obtained from the experiments in **b**, are given on the *ordinate* (logarithmic) scale: the abscissa gives the time of incubation with 10 m*M* carbamoylcholine before influx was assayed [see Eq. (7)]. The coordinates of the line were computed by use of Eq. (7). *Vertical bars* give the standard deviations. The *ordinate intercept* gives a value for $(J_I)_{T=0}$ of $1.3 \pm 0.2\ s^{-1}$, and the *slope* gives β as $0.12 \pm 0.03\ s^{-1}$, $\chi^2 = 0.8$ (*Bevington* 1969). (*Hess* et al. 1982)

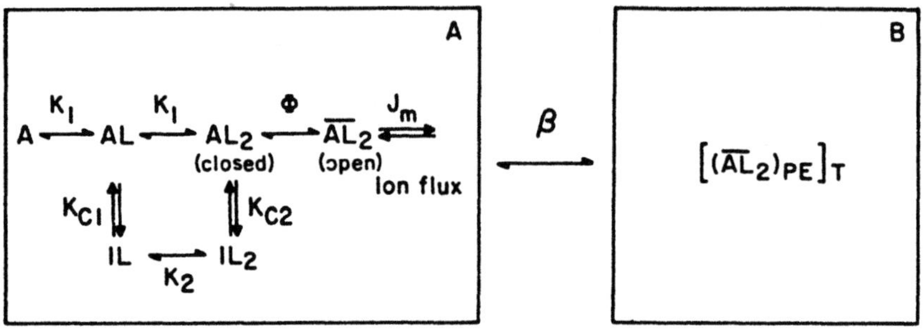

$$\frac{M_t}{M_\infty} = 1 - \exp - \bar{J} R \left\{ \left[(\overline{AL}_2)_0 - (\overline{AL}_2)_\infty \right] \frac{1-e^{-\alpha t}}{\alpha} + (\overline{AL}_2)_\infty t \right\} \quad [1]$$

$$\frac{M_t}{M_\infty} = 1 - \exp - \bar{J} R \left\{ \left[(\overline{AL}_2)_0 - (\overline{AL}_2)_{PE} \right] \frac{1-e^{-\alpha t}}{\alpha} + (\overline{AL}_2)_{PE} \frac{1-e^{-\beta t}}{\beta} \right\} \quad [2]$$

$$(J_A)_T = (J_A)_{T=0}\, e^{-\alpha t} \quad [3]$$

$$(J_I)_T = (J_I)_{T=0}\, e^{-\beta t} \quad [4]$$

Fig. 18. The minimal mechanism to account for the rates of acetylcholine-receptor-controlled cation translocation in *T. californica* membrane vesicles is based on measurements with *E. electricus* vesicles (*left* part of the scheme). All the symbols and constants pertaining to this model have been defined in the text. On the *right* is an extension of the model (Fig. 15) to account for the measurements with *T. californica* vesicles. The major difference between the results obtained with the acetylcholine receptor from the two organisms is that a second inactivation step occurs in the case of *T. californica.* It is assumed that in this step the concentration of the open-channel form is reduced to a low final value which is no longer detected in the flux assay. The rate coefficient for this inactivation step is designated β. The integrated rate equation [Eq. (1) in Figure] pertaining to the model based on flux measurements with *E. electricus* vesicle preparations (Fig. 15) can be extended to accommodate a second inactivation reaction. When $\alpha \gg \beta$, $(\overline{AL}_2)_\infty$ [Eq. (1) in Figure] can be considered to be in preequilibrium with all the other acetylcholine receptor forms, we designate this species $(\overline{AL}_2)_{PE}$ and obtain Eq. (2) in the Figure, the integrated rate equation for *Torpedo* spp. acetylcholine-receptor-controlled ion flux. The values of α and β can be obtained by use of Eqs. (3) and (4) in the Figure [which are Eqs. (6) and (7) in the text]. The subscript T refers to the preincubation time. The subscripts t and ∞ refer to time and the end of the experiment, respectively. M and R represent concentrations of metal ion and receptor respectively. (*Hess* et al. 1982)

The lack of correlation between toxin site concentration and ion flux rates makes direct comparison of results obtained with preparations from different species tenuous. However, the ratio of constants obtained with the same preparation is expected to be independent of the value of R_0. For example, the ratio of the values of $(J_A)_{T=0}$ in 10 m*M* carbamoylcholine to $(J_A)_{T=0}$ in 1 m*M* carbamoylcholine is 4:1, the corresponding ratio in *E. electricus* is 5:1. This indicates that the constants for the binding of carbamoylcholine to the active form of the receptor are similar in the two organisms. The ratio of $(J_A)_{T=0}$ to $(J_I)_{T=0}$ in 10 m*M* carbamoylcholine is about 300:1 in *T. californica* and about 400:1 in

E. electricus. The inactivation rate coefficient, α, which reflects the isomerization rate constants, is 3.5 times higher in *E. electricus* than in *T. californica.* The major difference detected between the acetylcholine receptor in the electroplax vesicles prepared from *E. electricus* and from *T. californica* is an additional and much slower inactivation reaction in *T. californica* vesicles (*Walker* et al. 1981). This difference may reflect not a species difference but a difference between extrasynaptic and synaptic receptors. The receptors in *E. electricus* vesicles may come mainly from the extrasynaptic region of the conducting membranes, and those in the *Torpedo* spp. membrane vesicles used in ion flux studies may be from the postsynaptic region (*Changeux* 1981).

A minimum extension of the model derived from measurements with *E. electricus* electroplax membrane vesicles will also accommodate the ion translocation process in *Torpedo* spp. vesicles. This model, and the corresponding integrated rate equation for ion flux, are shown in Fig. 18. The model can be treated quantitatively if one uses caution. Because α and β may have different ligand dependencies, an interpretation of the influx measurements requires an independent knowledge of the constants. The pulsed quench flow technique and the approach described to determine J_A, J_I, α, and β may be capable of providing the desired information. It should be noted that the integrated rate equation pertaining to the mechanism is only valid under restricted conditions, for instance, $J_A \gg \alpha$, J_I and $\alpha \gg \beta$.

9 Conclusion

At present the proposed mechanism, which relates ligand-binding steps, channel-opening equilibria, and rates for interconversion between active and inactive receptor forms to ion translocation rates, is the simplest one, accommodating all the kinetic and equilibrium data available. In addition, chemical kinetics may give unambiguous information about the rate constants for formation of receptor:ligand complexes and the opening of the channel. Methods which allow one to make these measurements are not yet available. Statistical methods give complementary information about steps that are independent of ligand concentration, the single-channel conductance, and the lifetime of the open channel.

The techniques which have been developed may be useful to obtain more detailed information about the compounds which modify receptor function and about the many other important molecules which control the transfer of molecules across cell membranes. The ability to (a) vary solutions on both sides of the membrane; (2) know the solution compositions, receptor concentrations, and ligand concentrations; and (c) break down complex kinetic processes into steps which obey simple rate laws and in which direct relationships between rates and ligand concentrations can be observed are the advantages of the techniques developed. The ability to isolate and thereby concentrate specific receptor proteins which exist in low concentrations in the nervous system and to reincorporate these proteins into membrane vesicles (*Racker* 1970; *Kaback* 1970; *Epstein* and *Racker* 1978; *Huganir* et al. 1979) should allow one to study many of the different receptors of the nervous system and the many other interesting proteins and enzymes which have been identified and which are involved in the transfer of molecules and ions across cell membranes.

Acknowledgements. The review was written while GPH was at the University of Konstanz, Germany, as a recipient of a Senior Scientist Award from the von Humboldt Foundation. We are grateful to Professor Peter Läuger, University of Konstanz, for helpful discussions. We are also grateful to Lisa Lapish who made the excellent membrane preparations with which most of the experiments were done, to Inge Schappeler and Melissa Stuckey for typing the manuscript, and to Carl Scriber for drawing the Figures. Finally we are grateful to all the members of the Laboratory whose work has been described here. Financial support from the National Institutes of Health and the National Science Foundation made the experiments possible.

References

Adams PR (1981) Acetylcholine receptor kinetics. J Membr Biol 58:161–174

Anderson CR, Stevens CF (1973) Voltage clamp analysis of acetylcholine-produced end-plate current fluctuations at frog neuromuscular junction. J Physiol (Lond) 235:655–691

Aoshima H, Cash DJ, Hess GP (1980) Acetylcholine-receptor-controlled ion flux in electroplax membrane vesicles. A minimal mechanism based on rate measurements in the millisecond to minute time region. Biochem Biophys Res Commun 92:896–904

Aoshima H, Cash DJ, Hess GP (1981) The mechanism of the inactivation (desensitization) of the acetylcholine receptor. Investigations by fast reaction techniques with membrane vesicles. Biochemistry 20:3467–3474

Bangham AD, Standish MM, Watkins JC (1965) Diffusion of univalent ions across the lamellae of swollen phospholipids. J Mol Biol 13:238–252

Bernhardt J, Neumann E (1978) Kinetic analysis of receptor-controlled tracer efflux from sealed membrane fragments. Proc Natl Acad Sci USA 75:3756–3760

Bernhardt J, Neumann E (1980) Physical factors determining gated flux from or into sealed membrane fragments. Neurochemistry Int 2:243–250

Bevington PR (1969) Data reduction and error analysis. Academic, New York

Bulger JE, Fu J-JL, Hindy EF, Silberstein RL, Hess GP (1977) Allosteric interactions between the membrane-bound acetylcholine receptor and chemical mediators. Kinetic studies. Biochemistry 16:684–692

Cash DJ, Hess GP (1980) Molecular mechanism of acetylcholine-receptor-controlled ion translocation across cell membranes. Proc Natl Acad Sci USA 77:842–846

Cash DJ, Hess GP (1981) Quenched flow technique with plasma membrane vesicles: acetylcholine-receptor-mediated transmembrane ion flux. Anal Biochem 112:39–51

Cash DJ, Aoshima H, Hess GP (1980) Acetylcholine-induced cation translocation across cell membranes and inactivation of the acetylcholine receptor: chemical kinetic measurements in the msec time region. Proc Natl Acad Sci USA 78:3318–3322

Cash DJ, Aoshima H, Hess GP (1981) Acetylcholine-induced receptor-controlled ion flux investigated by flow quench techniques. Biochem Biophys Res Commun 95: 1010–1016

Chance B, Eisenhardt RH, Gibson QH, Lonberg-Holm KK (eds) (1964) Rapid mixing and sampling techniques in biochemistry. Academic, New York

Changeux J-P (1981) The acetylcholine receptor: an "allosteric" membrane protein. Harvey Lect 75:85–254

Delgeane A, McNamee MB (1980) Independent activation of the acetylcholine receptor from *Torpedo californica* at two sites. Biochemistry 19:890–895

Epstein N, Racker E (1978) Reconstitution of carbamylcholine-dependent sodium ion flux and desensitization of the acetylcholine receptor from *Torpedo californica.* J Biol Chem 253:6660–6662

Epstein N, Hess GP, Kim PS, Noble RL (1980) Inactivation (desensitization) of the acetylcholine receptor in *Electrophorus electricus* membrane vesicles by carbamoylcholine: comparison between ion flux and α-bungarotoxin binding. J Membr Biol 56:133–137

Fersht AR, Jakes R (1975) Demonstration of two reaction pathways for the aminoacylation of tRNA. Application of the pulsed quench flow technique. Biochemistry 14:3350–3362

Froehlich JP, Taylor EW (1976) Transient kinetic effects of calcium ion on sarcoplasmic reticulum adenosine triphosphatase. J Biol Chem 251:2307–2315

Fu J-JL, Donner DB, Hess GP (1974) Half of the sites reactivity of the membrane-bound *Electrophorus electricus* acetylcholine receptor. Biochem Biophys Res Comm 60:1072–1080

Fu J-JL, Donner DB, Moore DE, Hess GP (1977) Allosteric interactions between the membrane-bound acetylcholine receptor and chemical mediators: equilibrium measurements. Biochemistry 16:678–684

Goldin SM, Rhodin V, Hess EJ (1980) Molecular characterization, reconstitution and "transport-specific fractionation" of the saxitoxin-binding protein/Na^+ gate of mammalian brain. Proc Natl Acad Sci USA 77:6884–6888

Grünhagen HH, Changeux J-P (1976) Studies on the electrogenic action of acetylcholine with *Torpedo marmorata* electric organ. J Mol Biol 106:517–535

Grünhagen HH, Iwatsubo M, Changeux J-P (1977) Fast kinetic studies on the interaction of cholinergic agonists with the membrane-bound acetylcholine receptor from *Torpedo marmorata* as revealed by quinacrine fluorescence. Eur J Biochem 80:225–242

Hammes GG (1978) Principles of chemical kinetics. Academic, New York

Hammes GG, Wu CW (1974) Kinetics of allosteric enzymes. Annu Rev Biophys Bioeng 3:1–33

Hazelbauer GH, Changeux J-P (1974) Reconstitution of a chemically excitable membrane. Proc Natl Acad Sci USA 71:1479–1483

Heidmann T, Changeux J-P (1978) Structural and functional properties of the acetylcholine receptor protein in its purified and membrane-bound states. Annu Rev Biochem 47:317–357

Heidmann T, Changeux J-P (1980) Interaction of a fluorescent agonist with the membrane-bound acetylcholine receptor from *Torpedo marmorata* in the millisecond time range: resolution and evidence for positive cooperative effects. Biochem Biophys Res Comm 97:889–896

Hess GP (1979) Acetylcholine-receptor-controlled ion fluxes in microsacs (membrane vesicles) obtained from the electroplax of *Electrophorus electricus*. In: Smith FO, Worden FG (eds) The neurosciences, fourth study program. MIT Press, Cambridge, pp 831–839

Hess GP, Andrews JP (1977) Functional acetylcholine receptor electroplax membrane microsacs (vesicles): purification and characterization. Proc Natl Acad Sci USA 74: 482–486

Hess GP, Andrews JP, Struve GE, Coombs SE (1975a) Acetylcholine-receptor-mediated ion flux in electroplax membrane preparations. Proc Natl Acad Sci USA 72:4371–4375

Hess GP, Bulger JE, Fu J-JL, Hindy EF, Silberstein RJ (1975b) Allosteric interactions of the membrane-bound acetylcholine receptor: kinetic studies with α-bungarotoxin. Biochem Biophys Res Comm 64:1018–1026

Hess GP, Andrews JP, Struve GE (1976) Apparent cooperative effects in acetylcholine-receptor-mediated ion flux in electroplax membrane preparations. Biochem Biophys Res Commun 69:830–837

Hess GP, Lipkowitz S, Struve GE (1978) Acetylcholine-receptor-mediated ion flux in electroplax membrane microsacs (vesicles): change in mechanism produced by asymmetrical distribution of sodium and potassium ions. Proc Natl Acad Sci USA 75:1703–1707

Hess GP, Cash DJ, Aoshima H (1979) Acetylcholine-receptor-controlled ion fluxes in membrane vesicles investigated by fast reaction techniques. Nature 282:329–331

Hess GP, Cash DJ, Aoshima H (1980) Kinetic mechanism of acetylcholine-receptor-controlled ion flux: flow quench kinetic measurements of acetylcholine-induced flux in membrane vesicles. Neurochemistry Int 2:233–242. Also in: Schoffeniels E, Neuman E (eds) (1981) Molecular aspects of bioelectricity. Pergamon, Oxford, p 233

Hess GP, Aoshima H, Cash DJ, Lenchitz B (1981) The specific reaction rate of acetylcholine-receptor-controlled ion translocation: a comparison of measurements with membrane vesicles and with muscle cells. Proc Natl Acad Sci USA 78:1361–1365

Hess GP, Pasquale EB, Karpen JW, Sachs AB, Takeyasu K, Cash DJ (1982a) Acetylcholine receptor-controlled ion translocation. A comparison of the effects of suberyldicholine, carbamoylcholine, and acetylcholine. Biochem Biophys Res Commun 107:1583–1588

Hess GP, Pasquale EB, Walker JW, McNamee MG (1982b) Comparison of acetylcholine-receptor-controlled cation flux in membrane vesicles from *Torpedo californica* and *Electrophorus electricus:* chemical kinetic measurements in the millisecond region. Proc Natl Acad Sci USA 79:963–967

Hess GP, Cash DJ, Aoshima H (1983) Acetylcholine-receptor-controlled ion translocation. Chemical kinetic investigations of the mechanism. Annu Rev Biophys Bioeng 12:443–473

Hess GP, Kolb H-A, Läuger P, Schoffeniels E, Schwarze W (1984) Acetylcholine receptor (from *E. electricus*): a comparison of single-channel current recordings and chemical kinetic measurements. Proc Natl Acad Sci USA (in press)

Huganir RL, Schell MA, Racker E (1979) Reconstitution of the purified acetylcholine receptor from *Torpedo californica*. FEBS Lett 108:155–160

Jürss R, Prinz H, Maelicke A (1979) NBD-5-acetylcholine: fluorescent analog of acetylcholine and agonist at the neuromuscular junction. Proc Natl Acad Sci USA 76:1064–1068

Kaback HR (1970) Transport. Annu Rev Biochem 39:561–598

Kandel ER (1981) Calcium and the control of synaptic strength by learning. Nature 293:697–700

Karlin A (1980) Molecular properties of nicotinic acetylcholine receptors. In: Cotman CW, Pate G, Nicolsen GL (eds) The cell surface and neuronal function. Elsevier/North Holland, Amsterdam, pp 191–260

Kasai M, Changeux J-P (1971a) In vitro excitation of purified membrane fragments by cholinergic agonists. I. Pharmacological properties of the excitable membrane fragments. J Membr Biol 6:1–23

Kasai M, Changeux J-P (1971b) In vitro excitation of purified membrane fragments by cholinergic agonists. II. The permeability change caused by cholinergic agonists. J Membr Biol 6:24–57

Kasai M, Changeux J-P (1971c) In vitro excitation of purified membrane fragments by cholinergic agonists. III. Comparison of the dose-response curves to decamethonium with the corresponding binding curves of decamethonium to the cholinergic receptor. J Membr Biol 6:58–80

Katz B (1966) Nerve, muscle, and synapse. McGraw-Hill, New York

Katz B (1969) The release of neural transmitter substances. Liverpool University Press, Liverpool

Katz B, Miledi R (1970) Membrane noise produced by acetylcholine. Nature 226: 962–963

Katz B, Miledi R (1972) The statistical nature of the acetylcholine potential and its molecular components. J Physiol (Lond) 224:665–669

Katz B, Miledi R (1977) Transmitter leakage from motor nerve endings. Proc R Soc Lond (Biol) 196:59–72

Katz B, Thesleff S (1957) A study of the desensitization produced by acetylcholine at the motor end-plate. J Physiol (Lond) 138:63–80

Keynes RD, Martins-Ferreira H (1953) Membrane potentials in the electroplates of the electric eel. J Physiol (Lond) 119:315–351

Kim PS (1980) Acetylcholine-receptor-rich electroplax membrane vesicles: identification and characterization of membrane properties that interfere with measurements of receptor-mediated ion flux. AB Honors Thesis, Cornell University

Kim PS, Hess GP (1981) Acetylcholine-receptor-controlled ion flux in electroplax membrane vesicles: identification and characterization of membrane properties that affect ion flux measurements. J Membr Biol 58:203–211

Kurzmack M, Verjovski-Almeida S, Inesi G (1977) Detection of an initial burst of Ca^{2+} translocation in sarcoplasma reticulum. Biochem Biophys Res Commun 78:772–776

Lehninger AL (1964) The mitochondrion. Benjamin, New York

Lee CY (1972) Chemistry and pharmacology of polypeptide toxins in snake venoms. Annu Rev Pharmacol 12:265–286

Leprince P (1981) Characterization of the reaction of α-bungarotoxin with acetylcholine receptors in *Torpedo californica* and *Electrophorus electricus* membrane preparations. MS Thesis, Cornell University

Leprince P, Noble RL, Hess GP (1981) Comparison of the interactions of a specific neurotoxin (α-bungarotoxin) with the acetylcholine receptor in *Torpedo californica* and *Electrophorus electricus* membrane preparations. Biochemistry 20:5565–5570

Lindstrom J, Patrick J (1974) Purification of the acetylcholine receptor by affinity chromatography. In: Synaptic transmission and nerve interaction. Raven, New York, pp 191–216

Lymn RW, Taylor EW (1970) Transient state phosphate production in the hydrolysis of nucleoside triphosphate by myosin. Biochemistry 9:2975–2983

Martonosi A, Lawinska E, Oliver M (1974) Elementary processes in the hydrolysis of ATP by sarcoplasmic reticulum membranes. Ann NY Acad Sci 227:549–567

McNamee MG, McConnell HM (1973) Transmembrane potentials and phospholipid flip-flop in excitable membrane vesicles. Biochemistry 12:2951–2958

Miller DL, Moore H-PH, Hartig PR, Raftery MA (1978) Fast cation flux from *Torpedo californica* membrane preparations: implications for a functional role for acetylcholine receptor dimers. Biochem Biophys Res Comm 85:632–640

Moore H-PH, Raftery MA (1980) Direct spectroscopic studies of cation translocation by *Torpedo* acetylcholine receptor on a time scale of physiological relevance. Proc Natl Acad Sci USA 77:4509–4513

Nachmansohn D (1955) Metabolism and function of the nerve cell. Harvey Lect 49: 57–99

Nachmansohn D, Neumann E (1975) Chemical and molecular basis of nerve activity. Academic, New York

Neher E, Sakmann B (1976) Single-channel currents recorded from membrane of denervated frog muscle fibres. Nature 260:779–802

Neher E, Stevens CF (1977) Conductance fluctuations and ionic pores in membranes. Annu Rev Biophys Bioeng 6:345–381

Neubig RR, Cohen JB (1980) Permeability control by cholinergic receptors in *Torpedo* postsynaptic membranes: agonist dose-response relations measured at second and millisecond times. Biochemistry 19:2770–2779

Neubig RR, Boyd ND, Cohen JB (1982) Conformations of *Torpedo* acetylcholine receptor associated with ion transport and desensitization. Biochemistry 21:3460–3467

Passow H (1969) Ion permeability of erythrocyte ghosts. In: Passow H, Stampli R (eds) Laboratory techniques in membrane biophysics. Springer, Berlin Heidelberg New York, p 21

Podleski TR, Bartels E (1963) Difference between tetracaine and *d*-tubocurarine in the competition with carbamoylcholine. Biochem Biophys Acta 75:387–396

Popot JL, Sugiyama H, Changeux J-P (1974) Démonstration de la sensibilisation pharmacologique des récepturs de l'acétylcholine in vitro avec des fragments de membrane excitable de torpille. CR Acad Sci (D) Paris 279:1721–1724

Popot J-L, Sugiyama H, Changeux J-P (1976) Studies on the electrogenic action of acetylcholine with *Torpedo marmorata* electric organ. II. The permeability response of the receptor-rich membrane fragments to cholinergic agonists in vitro. J Mol Biol 106:469–483

Racker E (ed) (1970) Membranes of mitochondria and chloroplasts. Van Nostrand-Reinhold, New York

Ramseyer GO, Morrison GH, Aoshima H, Hess GP (1981) Vidicon flame emission spectroscopy of Li^+, Na^+, and K^+ fluxes in *Electrophorus electricus* membrane vesicles. Anal Biochem 115:34–46

Rang HP (1974) Acetylcholine receptors. Q Rev Biophys 7:283–399

Roughton FJW, Chance B (1963) In: Friess SL, Lewis ES, Weissberger A (eds) Technique of organic chemistry, 2nd edn. Wiley, New York

Sachs AB (1982) The acetylcholine receptor: characterization of α-bungarotoxin sites in *Electrophorus electricus* membrane preparations, development of fluorescent quenching techniques to measure ion flux, and the effects of phencyclidine on the receptor in living nerve cells. AB Honors Thesis, Cornell University

Sachs AB, Lenchitz B, Noble RL, Hess GP (1982) A new method for large scale preparation of membrane vesicles which are selectively permeable to specific ions: acetylcholine-receptor-containing vesicles. Anal Biochem 124:185–190

Sakmann B, Adams PR (1978) In: Jacob J (ed) Advances pharmacology and therapeutics. Pergamon, Oxford, pp 81–90

Sakmann B, Patlack J, Neher E (1980) Single acetylcholine-activated channels show burst kinetics in the presence of desensitizing agonist concentrations. Nature 286: 71–73

Schoffeniels E, Nachmansohn (1957) An isolated electroplax preparation. II. Improved preparation for studying ion flux. Biochim Biophys Acta 26:585–596

Sheridan RE, Lester HA (1977) Rates and equilibria at the acetylcholine receptor of *Electrophorus* electroplax. A study of neurally evoked postsynaptic currents and of voltage-jump relaxations. J Gen Physiol 70:187–219

Sugiyama H, Popot JL, Cohen JB, Weber M, Changeux J-P (1975) In: Sund H, Blauer G (eds) Protein-ligand interactions. deGruyter, Berlin, pp 289–305

Sumida M, Wang T, Mandel F, Froehlich JP, Schwartz A (1978) Transient kinetics of Ca^{2+} transport of sarcoplasmic reticulum. J Biol Chem 253:8772–8777

Thayer WS, Hinkle PC (1975) Kinetics of adenosine triphosphate synthesis in bovine heart submitochondrial particles. J Biol Chem 250:5336–5342

Toor HL (1975) The non-premixed reaction A+B products. In: Brodkey RS (ed) Turbulence in mixing operations. Academic, New York, p 121

Verjovski-Almeida, Inesi G (1979) Fast kinetic evidence for an activating effect of ATP on the Ca^{2+} transport sarcoplasmic reticulum ATPase. J Biol Chem 254:18–21

Walker JW, McNamee MG, Pasquale E, Cash DJ, Hess GP (1981) Acetylcholine receptor inactivation in *Torpedo californica* electroplax membrane vesicles. Detection of two processes in the millisecond and second time regions. Biochem Biophys Res Comm 100:86–90

Rev. Physiol. Biochem. Pharmacol., Vol. 102
© by Springer-Verlag 1985

Organization of the Lumbar Sympathetic Outflow to Skeletal Muscle and Skin of the Cat Hindlimb and Tail

WILFRID JÄNIG

Contents

Physiologisches Institut der *Universität* Kiel, Olshausenstr. 40–60, 2300 Kiel, FRG

List of Abbreviations

BP	blood pressure
Co	coccygeal ganglion
CVC	cutaneous vasoconstrictor
CVD	cutaneous vasodilator
GS	nerves to medial gastrocnemius and lateral gastrocnemius-soleus muscles
HRP	horseradish peroxidase
imp/s	impulses per second
$L_1 \ldots L_7$	lumbar ganglia of the sympathetic trunk
LST	lumbar sympathetic trunk
MP	medial plantar nerve
m/s	meters per second
MVC	muscle vasoconstrictor
MVD	muscle vasodilator
PM	pilomotor
$S_1 \ldots S_5$	sacral ganglia of the sympathetic trunk
SM	sudomotor
SP	superficial peroneal nerve
SU	sural nerve
$\bar{x} \pm SD$	mean ± standard deviation
$\bar{x} \pm SE$	mean ± standard error
WR	white ramus

1 Introduction

1.1 General Background

The concept that motoneurones are the "final common motor paths" (*Sherrington* 1906), which are shared by spinal reflex pathways and descending motor activities, has dominated the analytical approach to the motor system for a long time. With the advent of more refined techniques of extra- and intracellular recording from interneurones and motoneurones, the notion of the "functional motor unit" of the spinal cord came more and more into focus. Functional motor units comprise pools of different types of interneurones, their synaptic interconnections and their different inputs; they are elementary motor programs, do not operate independently and are used by the supraspinal brain structures in conjunction with the primary afferent input systems in order to control the final common pathways. Spinal reflex pathways and descending motor pathways do not act in parallel, with respect to the motoneurones, but in a concerted manner in the creation of various types of movements (*Baldissera* et al. 1981; *Jankowska* and *Lundberg* 1981). Neurobiological work on the motor systems in vertebrates and also in invertebrates (*Kandel* 1976; *Muller* et al. 1981) has paradigmatically shown that thorough analysis of the synaptic connections between primary afferents, descending systems, interneurones and motoneurones is the necessary base for the creation of new hypotheses. These hypotheses can mostly only be tested during ongoing regulations of final common motor pathways, i.e. during simple and complex motor activities.

Analysis of the neuronal organization of the sympathetic nervous system (and of the autonomic nervous system in general) has not attained the stage at which the analysis of the motor system finds itself at the moment. This may have several reasons:

1. The sympathetic nervous system has been regarded as a unitary system from the functional point of view (*Cannon* 1929, 1939). Concepts about the functioning of the sympathetic nervous system are largely derived from experiments in which autonomic effector responses or multi-unit activity in sympathetic nerves have been recorded. The approaches have given valuable insight into the overall capacity of the system to regulate various bodily functions for maintenance of homeostasis, adaptability of the inner milieu and integration of the sympathetic actions during motor performances of the organism (*Cannon* 1929, 1939; *Folkow* and *Neil* 1971; *Koizumi* and *Brooks* 1972; *Korner* 1979). These approaches have given only minor insights into the central neuronal machinery, which enables the regulation of the autonomic effector organs by the sympathetic systems for maintaining homeostasis and adaptability.

2. Somata and axons of pre- and postganglionic sympathetic neurones are small, and isolated recordings from them in animals proved difficult. Technical difficulties were overcome recently, and systematic studies of single units in vivo were started some 15 years ago. This recent analytical neurophysiological approach, in conjunction with neurohistological, neuropharmacological and biochemical methods, should render new results and give deeper insights into the central neuronal circuits of the sympathetic system.

3. The peripheral sympathetic nervous system consists of pre- and postganglionic neurones arranged in series. This arrangement impedes the analysis of discharge patterns of preganglionic neurones with respect to different target organs, i.e. with respect to function. Furthermore, it impedes the elaboration of the size of the peripheral neural units in the sympathetic nervous system, in particular since many different target organs are innervated by the sympathetic nervous system.

4. The view of the sympathetic nervous system as being a unitary system and the coordinate smooth ongoing regulations of the autonomic target organs by this system may have impeded the systematic application of the reflex concept in the analysis of the sympathetic nervous system. This Sherringtonian approach has been neglected in the analysis of the sympathetic nervous system; many scientists working in this field have only limited confidence in the value of this approach when applied to the sympathetic nervous system (*Hilton* and *Spyer* 1980; *Koepchen* et al. 1980). However, this approach may give valuable information on the functional structure of the sympathetic nervous system which is indispensable for understanding the regulation of target organs by the sympathetic nervous system in the living non-anaesthetized organism. After all, nobody would doubt how prolific the application of this Sherringtonian concept was, and still is, in the analysis of the organization of the neuronal circuits in the spinal cord and in supraspinal brain structures, not withstanding that we still do not fully understand how locomotion and other coordinated movements in vertebrates are brought about (*Baldissera* et al. 1981; *Grillner* 1975, 1981).

Going back to the beginning of this introductory section, a similar analysis as that applied to the motor system should be used in an analytical approach to the sympathetic nervous system in order to illuminate the building blocks of this system. This analysis should start right from the beginning and should define the functional characteristics of the final common pathways and of the more complex units of this system.

1.2 Special Background

Some 15 years ago a systematic analysis of single pre- and postganglionic sympathetic units which were isolated from the cervical sympathetic trunk and from peripheral nerves supplying the cat hindlimb was started. The experiments concentrated on the analysis of somatosympathetic reflexes using electrical stimulation of group I to IV afferent fibres in skin and muscle nerves (*Sato* and *Schmidt* 1973). The aim of these experiments was to gain deeper insight into the functional organization of the sympathetic nervous system, both peripherally and centrally. It soon became obvious that this single-unit analysis of pre- and postganglionic neurones could render only limited insight into the neuronal organization of the sympathetic nervous system. For example, single preganglionic neurones projecting into the cervical sympathetic trunk (*Jänig* and *Schmidt* 1970) and postganglionic neurones projecting to the hindlimb (*Jänig* et al. 1972) exhibited a wealth of excitatory and inhibitory reflexes upon electrical stimulation of peripheral skin and muscle nerves. However, it was not possible to bring order into these reflexes and present a consistent picture of their organization, nor was it possible to make sense of the different types of reflexes elicited by electrical stimulation of myelinated and unmyelinated afferents via spinal and supraspinal integration levels (*Sato* and *Schmidt* 1973).

One way out of this dilemma was to concentrate on the functional identification of different types of postganglionic neurones and to use physiological stimulation of different afferent neurones. Initially, this approach was attempted successfully in a series of experiments which concentrated mainly on somatosympathetic reflexes in vasoconstrictor neurones (*Horeyseck* and *Jänig* 1974a,b,c; *Jänig* 1975; *Horeyseck* et al. 1972, 1976). In succeeding studies, the somatosympathetic reflexes receded more and more into the background. However, these reflexes still play an integral part in the functional identification of post- and preganglionic neurones supplying skeletal muscle and skin and also in conceptual thinking about the organization of the sympathetic system in the neuraxis.

1.3 Aims

This review will concentrate on the description of the sympathetic pre-postganglionic channels supplying the skin and muscle of the cat hindlimb and tail. Emphasis has been placed on the functional properties of both pre- and postganglionic neurones of these sympathetic pre-postganglionic channels. The description is based largely on experiments conducted over the last 10 years in the author's laboratory and therefore gives a personal view, which has evolved over the years, on the structure and function of the sympathetic system. It is not intended to cover the abundant literature which deals with the structure and function of sympathetic pre- and postganglionic neurones supplying skin and skeletal muscle. In particular, it will not cover investigations performed with indirect methods (analyses of effector responses) or studies in contiguous fields (e.g., circulation research, thermoregulation and morphology, pharmacology and development of sympathetic neurones). In these cases, if necessary, reference will be made to the respective reviews published recently.

The treatise is centred around two complementary problems:

1. The pre- postganglionic channels will be characterized functionally by the discharge patterns which can be elicited in the neurones by stimulation of receptors on the body surface and in the interior of the body. The separateness of these channels will further be corroborated by the descriptions of other functional properties of the neurones.

2. The discharge properties of the different sympathetic neurones will give valuable insight into the global organization of different sympathetic systems in the spinal cord and brainstem. Here, the comparison of results from brain-intact and spinal animals will be particularly valuable. In analogy to the functional motor units in the spinal cord (*Baldissera* et al. 1981), the notion of "spinal sympathetic functional units" will be introduced. It will be shown that the distinct reflex patterns of different types of sympathetic neurones could encourage further analysis of the neuronal organization of the sympathetic system in the spinal cord (i.e. its preganglionic neurones, interneurones and descending control systems).

2 Possible Types of Postganglionic Neurones Supplying Hindlimb and Tail and Methodological Considerations

2.1 Types of Postganglionic Neurones

Postganglionic neurones supplying the skeletal muscle and skin of the cat hindlimb and tail innervate various types of target organs. Six types of postganglionic neurones which can be identified neurophysiologically will be described (Fig. 1):

1. Postganglionic adrenergic vasoconstrictor neurones supplying skeletal muscle (MVC).

2. Postganglionic adrenergic vasoconstrictor neurones supplying hairy and hairless skin (CVC).

3. Postganglionic cholinergic sudomotor (SM) neurones supplying the sweat glands of the hairless skin. The hairy skin of the cat does not contain

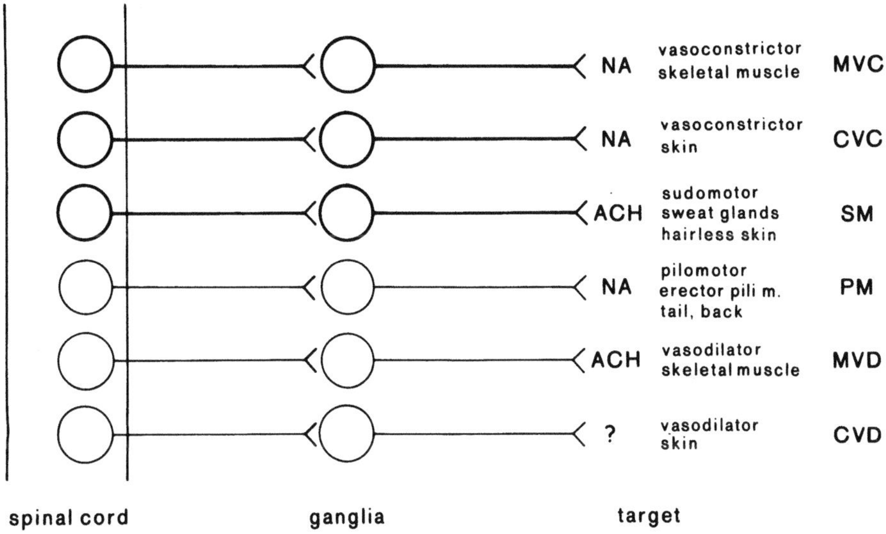

Fig. 1. Sympathetic systems supplying skeletal muscle and skin. The first three systems have ongoing activity. *NA*, noradrenaline; *ACH*, acetylcholine. (*Jänig* et al. 1983)

sweat glands (*Langley* 1891a, 1894/5); therefore, SM axons should exist only in nerves which supply the paws.

4. Postganglionic adrenergic pilomotor (PM) neurones supplying the erector pili muscles of hair follicles. Repetitive electrical stimulation of the LST leads to piloerection only on the tail and in the dorsal skin of the trunk. No piloerection is observed in the hairy skin of the hindlimb, except in a small dorsal skin area close to the tail. Therefore, it is very likely that skin nerves to the hindlimb do not contain PM axons whereas those to tail and back do (*Langley* and *Sherrington* 1891; *Langley* 1894; *Lewis* and *Marvin* 1927), although hair follicles of the hindlimb also seem to have small erector pili muscles (*Strickland* and *Calhoun* 1963).

5. Postganglionic cholinergic vasodilator neurones supplying the vessels of skeletal muscle (MVD). The existence of this type of vasodilator neurone in the cat has been well established by indirect methods (measurement of flow through skeletal muscle) (*Uvnäs* 1954, 1967; *Folkow* 1955; *Lisander* 1970; *Folkow* and *Neil* 1971). These neurones are activated during defence behaviour (defence reaction) and during electrical stimulation of the defence area in the hypothalamus.

6. Postganglionic vasodilator neurones supplying skin (CVD). The existence of this type of postganglionic neurone in carnivores has not been very well established. Its transmitter is unknown and its function is unclear (*Bell* 1983).

These six functional types of postganglionic neurones will be discussed in the following review. For the identification and functional characteriza-

Table 1. Types of postganglionic neurones supplying target organs of skin and skeletal muscle through different nerves

Nerve	Target organ	Type of postgang-lionic neurone	Response to stimulation
SU, SP (without branches to skeletal muscle)	Vascular bed of hairy skin	CVC CVD	Vasoconstriction Vasodilation
Skin nerves to tail (branches of caudo-ventral nerves)	Vascular bed of hairy skin, erector pili muscles	CVC CVD PM	Vasoconstriction Vasodilation Piloerection
Fascicles of MP to central pad of paw	Vascular bed of hairless skin, sweat glands	CVC CVD SM	Vasoconstriction Vasodilation Sweat production
Branches of deep peroneal nerve to anterior tibial muscle and long extensor muscle of toes, nerves to gastrocnemius-soleus muscle	Vascular bed of skeletal muscle	MVC MVD	Vasoconstriction Vasodilation

tion of these types of neurones, hypotheses were used which had developed from the physiology of regulation of various body functions (e.g. temperature regulation, regulation of cardiovascular system, regulation of defence behaviour). A seventh type of postganglionic neurone, which may be involved in efferent sympathetic control of receptors, will also be briefly discussed since it is of general interest. The experimental evidence for the individuality of this type of postganglionic neurone is very limited (see Sect. 3.7). Further separate types of postganglionic neurones supplying skin and skeletal muscle may exist (for example, postganglionic neurones supplying the fat cells in the subcutaneous tissue: see *Rosell* and *Belfrage* 1979); however, at present we have no direct experimental access to them since we do not know how to activate them selectively.

Various nerves to the skin and skeletal muscle of the hindlimb and tail contain different compositions of postganglionic axons. This is summarized in Table 1 for those five nerves which were used in the single-unit analysis of postganglionic neurones.

2.2 Methodological Considerations

Before going into details of a description of the discharge patterns in sympathetic neurones, it is important to note that most experiments were conducted under standardized conditions. Though self-evident, these experimental conditions must always be kept in mind when comparisons are made between results obtained on post- and preganglionic neurones and those obtained on circulation parameters etc. under unanaesthetized conditions (see *Jänig* et al. 1983).

The experimental conditions were as follows: chloralose anaesthesia (initially, 30–50 mg/kg i.v. or i.p.; during the experiments, doses of 10–30 mg i.v. in order to obtain miosis of the pupils) with or without an initial application of ketamine HCl (10–15 mg/kg i.m.); artificial ventilation with a positive pressure pump from the beginning of the surgery (end-expiratory CO_2 adjusted to 3–4 vol. %) with immobilization by gallamine triethiodide (Flaxedil, 10–20 mg/h) or by pancuronium bromide (about 0.6 mg/h); body core temperature 37.5°–38.5°C; mean arterial PB 100 mmHg or higher. The volume was infused i.v. (about 1–5 ml kg^{-1} h^{-1}; 100 ml solution was composed of 70 ml Tyrode's and 30 ml 10% glucose containing 3 mmol bicarbonate) according to the production of urine and to the extent of the surgery. When the arterial PB was below 90 mmHg, the characteristic reflexes mostly could no longer be elicited in the postganglionic neurones. Some experiments on the SM system were conducted under ketamine anaesthesia or under combinations of anaesthetics.

2.2.1 Recording of Activity in Post- and Preganglionic Neurones

For the recording of efferent activity, postganglionic axons were isolated in bundles from the respective peripheral nerves (see Table 1) and preganglionic axons in bundles from the LST distal to ganglion L_4. The arrangement of recording and stimulation electrodes is illustrated in Fig. 2A. Postganglionic activity was recorded from one or two bundles simultaneously (rec fibres in Fig. 2A). The preganglionic axons in the LST were stimulated between ganglia L_3 and L_4 or L_4 and L_5 (stim LST in Fig. 2A). These positions were preganglionic for at least 97% of the postganglionic neurones supplying the hindlimb and tail (*McLachlan* and *Jänig* 1983; see also *Langley* 1903). A second pair of electrodes for stimulating the postganglionic axons was positioned at the peripheral nerve (stim post in Fig. 2A). Figure 2B illustrates the electrophysiological identification of two postganglionic axons in a bundle with three unmyelinated axons (1, 2, 3) by colliding the impulses elicited from the peripheral nerve (stim post) with the impulses elicited from the preganglionic side in the LST (stim LST). Two axons were postganglionic (1, 2); one unmyelinated axon could not be activated from the LST and was most likely afferent (3). Preganglionic axons (rec pre in Fig. 2A) were identified by stimulation of the WR with single and repetitive impulses (stim WR in Fig. 2A; Fig. 2C) (see *Jänig* and *Szulczyk* 1980, 1981).

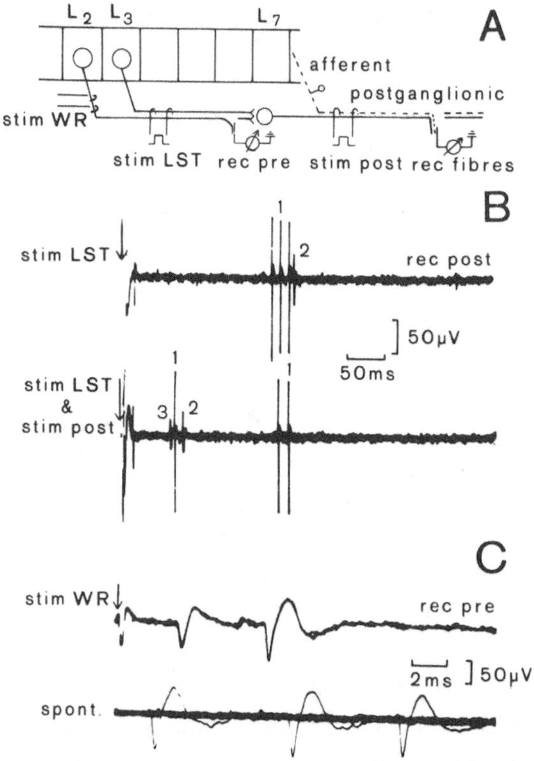

Fig. 2A–C. Identification of pre- and postganglionic neurones. **A** Arrangement of stimulation (stim WR, LST, post) and recording electrodes at the pre- and postganglionic sides (*rec pre, rec fibres*). **B** Recording from a bundle dissected from the SP with two postganglionic fibres (*1, 2*) and one (afferent) unmyelinated fibre (*3*), which could not be activated by stimulation of the LST (*stim LST*). *Upper trace,* stimulation of the LST (10 V); neurone *1* responded with three impulses, neurone *2* with one impulse. *Lower trace,* supramaximal stimulation of the LST and peripheral nerve (*stim post,* 20 V); direct excitation of fibres *1* and *2* and collision of first response elicited in fibre *1* and response elicited in fibre *2* from the LST. Additional excitation of fibre *3*. **C** Recording from a bundle with two axons dissected from the LST. Stimulation of the WR L_2 (*stim WR*). One axon had ongoing activity and therefore was preganglionic (see spont. *lower record*). (Modified from *Jänig* and *Szulczyk* 1981 and *Blumberg* and *Jänig* 1982)

2.2.2 Recording of Responses of Target Organs

Responses of target organs were recorded by indirect means, when necessary and when possible. The skin potential was recorded on the surface of the hairless skin (central pad) of hindpaws and forepaws. Fast negative deflections of the skin potential are preceded by discharges in postganglionic SM neurones (*Jänig* and *Räth* 1977, 1980; *Jänig* and *Kümmel* 1977, 1981). The skin temperature was recorded on the surface of the hairless skin (central pad) of the hindpaws. A decrease of activity in CVC neurones is followed by an increase in skin temperature, an increase of postganglionic activity by a decrease in skin temperature (*Jänig* and *Kümmel* 1981; *Grewe* et al. 1982). Contraction of erector pili muscles of the tail was observed by inspection of erection of hairs (*Grosse* and *Jänig* 1976). Vasodilation and vasoconstriction in skeletal muscles of the hindlimb were measured by recording blood flow changes through the descending aorta distal to the inferior mesenteric artery (*Horeyseck* et al. 1976).

2.2.3 Stimulation of Afferent Systems

Essential for the analysis of the discharge patterns in the post- and preganglionic neurones was the adequate stimulation of afferent systems.

1. The influence of arterial baroreceptor activity on postganglionic neurones was tested in two largely equivalent ways. First, the baroreceptors were stimulated by pressure increases applied to an isolated carotid sinus (carotid blind sac); the buffer nerve from the isolated carotid sinus was left intact, the other buffer nerves (contralateral carotid sinus nerve, both vagoaortic nerves) were cut. Second, the afferents from most arterial baroreceptors (in either both carotid sinus nerves or in the left carotid sinus nerve, and in both vagoaortic nerves) were left intact; in this way the baroreceptors were physiologically stimulated by the pulse BP (for literature see *Kirchheim* 1976). This leads to pulsatile modulation of activity in sympathetic vasoconstrictor neurones (cardiac rhythmicity). The degree of inhibition of postganglionic activity produced by a pressure step in an isolated carotid blind sac is highly correlated with the degree of cardiac rhythmicity of postganglionic activity (*Gregor* et al. 1977; *Blumberg* et al. 1980).

2. Arterial chemoreceptors were also stimulated in two ways: First, boluses of small volumes of CO_2-enriched Tyrode's solution were injected into the lingual artery. Second, the animal was ventilated with a hypoxic gas mixture of 8% O_2 in N_2 (very occasionally 5% O_2 in N_2) for 2–3 min. Mostly, both methods of stimulation produced equivalent reflexes in the vasoconstrictor neurones (*Gregor* and *Jänig* 1977; *Blumberg* et al. 1980).

3. Cutaneous nociceptors of hind- and forelimb were stimulated mechanically with toothed forceps, radiant heat or water at variable temperatures above 45°C (*Horeyseck* and *Jänig* 1974b; *Jänig* and *Räth* 1977; *Jänig* and *Spilok* 1978).

4. Afferents from hair follicles were stimulated by short air jets applied to the trunk and limbs (*Horeyseck* and *Jänig* 1974a).

5. Pacinian corpuscles in the paws were stimulated by vibration of the feet. The vibration was produced by tapping on the holders of the feet or the experimental frame to which the hindfeet were rigidly fixed (*Jänig* and *Räth* 1977; *Jänig* and *Kümmel* 1977).

6. Afferents from the urinary bladder and the colon were stimulated by increasing intraluminal pressures in the organs, thus distending their walls, and by isovolumetric contractions of the organs. These procedures excited visceral afferents in pelvic and hypogastric nerves which conveyed noxious as well as innoxious information (*Blumberg* et al. 1983).

7. Spinal thermoreceptors (or thermoreceptive neurones) were stimulated by warming and cooling the spinal cord. This was performed with a water-perfused U-shaped polyethylene tubing, which was positioned into the epidural space through an opening between the lumbar vertebra L_7 and the sacral bone (*Gregor* et al. 1976; *Jänig* and *Kümmel* 1981). Hypothalamic thermoreceptors (or thermoreceptive neurones) were stimulated by a water-perfused probe, which was positioned medially into the anterior hypothalamus (*Grewe* et al. 1982).

2.2.4 Morphological Investigations

In these investigations, HRP solution was applied to the skin and muscle nerves of the cat hindlimb and to the proximal end of the cut lumbar sympathetic chain between ganglia L_5 and L_6. The investigations were supplementary to the neurophysiological investigations and gave numbers and locations of post- and preganglionic neurones (*McLachlan* and *Jänig* 1983).

3 Reflex Patterns of Postganglionic Neurones in Brain-Intact Animals

For clarity, the different sympathetic systems in this section will be described separately. This approach is somewhat artificial and neglects the integrative point of view; however, it is the only way in which the functional characteristics of the systems can be extensively described. All properties of the sympathetic systems in animals with intact neuraxis are listed in Table 2a. It must be emphasized that each functional label given to a neurone, as obvious as the label may appear, derived from correlational studies and from the reflex (reaction) patterns. No direct experimental evidence exists which shows that a neurone being recorded from actually innervates a specific target organ (see Sect. 5.2).

3.1 Vasoconstrictor Neurones Supplying Skeletal Muscle (MVC)

3.1.1 Resting Activity

The rate of resting activity in MVC neurones is usually about 0.5–3 imp/s. This may vary considerably from cat to cat, seemingly under the same

Table 2a. Properties of sympathetic neurones: brain-intact animals

	MVC	CVC	SM	PM	MVD	CVD
1 Target organ	Vascular bed of skeletal muscle	Vascular bed of hairy and hairless skin	Sweat glands in hairless skin of paws	Erector pili muscles of the tail	Vascular bed of skeletal muscle	Vascular bed hairy and hairless skin
Transmitter	Noradrenaline	Noradrenaline	Acetylcholine	Noradrenaline	Acetylcholine	?
Effector response	Vasoconstriction	Vasoconstriction	Sweat secretion	Piloerection	Vasodilation	Vasodilation
2 *Conduction velocity of axons* (m/s, x ± SD)						
Preganglionic	3.4 ± 1.9 (*n*=59)	6.7 ± 2.9 (*n*=79) 1.6 ± 0.7 (*n*=18)	5–18 (*n*=6)	9.9 ± 2.9 (*n*=52)		
Postganglionic	0.79 ± 0.22 (*n*=55)	0.59 ± 0.13 (*n*=193)	0.81 ± 0.1 (*n*=48)	0.84 ± 0.14 (*n*=111)	Unmyelinated > MVC	Unmyelinated (? postganglionic)
3 *Ongoing activity* (imp/s, x ± SD)						
Preganglionic	1.8 ± 1.3 (*n*=26)	0.9 ± 0.6 (*n*=47)	Some	(Ø)	(Ø)	(Ø)
Postganglionic	−0.5–3	1.3 ± 0.9 (*n*=20) 1.1 ± 0.4 (*n*=24) 0.4–4	0.2 ± 0.1 (*n*=15) or silent	Silent	Silent	Silent
4 *Arterial baroreceptor input* Cardiac rhythm (%, $\overline{x}$ ± SE)						
Preganglionic	127 ± 12 (*n*=26)	23 ± 6 (*n*=43)	Ø	(Ø)	(Ø)	(Ø)
Postganglionic	131 ± 12 (*n*=33)	55 ± 6 (*n*=54) (hairy skin) 36 ± 6 (*n*=19) (hairless skin)	Ø	Ø	Ø	Ø

Table 2a (continued)

	MVC	CVC	SM	PM	MVD	CVD
Stimulation of baroreceptors in blind sac	↓	∅ or ↓ Some ↓				
5 *Chemoreceptor input*						
Direct stimulation	↑	↓, Some ↑	↑	∅	∅	∅
Systemic hypoxia (8% in N_2) ($\overline{x} \pm SE$)	↑ (3.3 ± 0.5 times)	Most ↓ (0.54 ± 0.08 times) Some ↑ (only hairy skin)	↑	↑ (asphyxia)		
Systemic hypercapnia (7% CO_2, 21% O_2 in N_2)	↑	Most ↓ Some ↑	↑			
6 *Thermal inputs*						
Spinal cord: warm	∅	↓	↑ or ∅			
Spinal cord: cold	∅	↑ or ∅	↑ or ∅			
Hypothalamus: warm	∅	↓	↑ or ∅			
Hypothalamus: cold	∅	↑ or ∅	↑ or ∅			
7 *Cutaneous inputs*						
Nociceptor (mech., heat)	↑	↓ Some ↑ (spatial org.)	↑	∅	∅	∅
Hair follicle receptors (air jet)	↓	↑	∅	∅	∅	∅
Pacinian corpuscles (paw)		↓ or ∅	↑			

8 *Visceral inputs* Nociceptive, bladder and colon contraction, distension	⬆	⬇	⬆	∅	∅
9 *Respiratory rhythmicity*	Strong	Weak, some strong	Weak, some strong		

⬆ ↑ large/small increase of activity, ⬇ ↓ large/small decrease of activity, ∅ no effect, (∅) probably no effect. To *2:* conduction velocities of preganglionic axons determined with direct and indirect method (see Fig. 15A) pooled together; conduction velocities of postganglionic axons corrected to about 33°C (see *Blumberg* and *Jänig* 1982). To *4:* cardiac rhythmicity of activity evaluated from post-R-wave histograms: activity in 48 ms of maximum minus activity in 48 ms minimum divided by mean activity in 48 ms times 100. To *5:* Reactions to systemic hypoxia expressed with respect to mean rate of ongoing activity.

Sources of Data.
For 2 Preganglionic. MVC, CVC, SM: *Jänig* and *Szulczyk* 1980; PM: *Grosse* and *Jänig* 1976.
For 2 Postganglionic. MVC: *Jänig* and *[illegible]elke*, unpublished; CVC: *Jänig* and *Szulczyk* 1981; SM: *Jänig* and *Kümmel* 1977; PM: *Grosse* and *Jänig* 1976; MVD: *Horeyseck* et al. 1976; CVD: *Jänig* and *Kümmel* 1981.
For 3 Preganglionic. MVC, CVC, SM: *Jänig* and *Szulczyk* 1980.
For 3 Postganglionic. MVC: *Jänig*, unpublished; CVC: *Grosse* and *Jänig* 1976, *Jänig* and *Kümmel* 1977; SM: *Jänig* and *Kümmel* 1977; PM: *Grosse* and *Jänig* 1976; CVD: *Gregor* et al. 1976, *Jänig* and *Kümmel* 1981; *MVD: Horeyseck* et al. 1976.
For 4 Preganglionic. MVC, CVC, SM: *Jänig* and *Szulczyk*, unpublished, *Jänig* 1984.
For 4 Postganglionic. MVC, CVC: *Blumberg* et al. 1980, *Gregor* and *Jänig* 1977, *Jänig* and *Kümmel* 1977; SM: *Jänig* and *Kümmel* 1977.
For 5. MVC, CVC: *Blumberg* et al. 1980, *Gregor* and *Jänig* 1977, *Jänig* and *Kümmel* 1977; SM: *Jänig* and *Kümmel* 1977, *Jänig* and *Kümmel* 1981; PM: *Grosse* and *Jänig* 1976.
For 6 Spinal cord. MVC: *Grewe*, unpublished; CVC, SM: *Gregor* et al. 1976, *Jänig* and *Kümmel* 1981.
For 6 Hypothalamus. MVC, CVC, SM: *Grewe* et al., unpublished, *Grewe* et al. 1982.
For 7 Nociceptor. MVC: *Horeyseck* and *Jänig* 1974b, *Jänig* and *Szulczyk* 1980; CVC: *Grosse* and *Jänig* 1976, *Horeyseck* and *Jänig* 1974b, *Jänig* and *Kümmel* 1977; SM: *Jänig* and *Kümmel* 1977, *Jänig* and *Räth* 1977.
For 7 Hair follicle. MVC, CVC: *Grosse* and *Jänig* 1976, *Horeyseck* and *Jänig* 1974a; SM: *Jänig* and *Kümmel* 1977.
For 7 Pacinian corpuscle. CVC, SM: *Jänig* and *Kümmel* 1977, *Jänig* and *Kümmel* 1981, *Jänig* and *Räth* 1977.
For 8. MVC, CVC, SM: *Blumberg* et al. 1983.
For 9. MVC, CVC: *Gregor* et al. 1977, *Jänig* et al. 1980; SM: *Jänig* and *Kümmel* 1977, *Jänig* et al. 1980

Table 2b. Properties of sympathetic neurones: spinal animals

	MVC	CVC	SM
1 Ongoing activity (imp/s)	0.5–1.5 or silent	0.1–1.5 (mean ∿ 0.5) or silent	0.2 ± 0.2 (*n* = 20) or silent
2 *Thermal inputs*			
Spinal cord: warm		↓	↑ or ∅
cold		↑	↑ or ∅
3 *Cutaneous inputs*			
Nociceptors (mech., heat)	↑	↓ (ipsilat., spatial org.), some ↑	↑
Hair follicle receptors	↓ or ∅	↓ or ∅	∅
Pacinian corpuscles (vibration)		∅, Some ↓	↑
4 *Visceral inputs*			
Nociceptive; bladder, colon, anus	↑	↑	↑

Symbols: see legend to Table 2a.

Sources of Data:
For 1. MVC: ***Horeyseck*** and ***Jänig*** 1974c; CVC: ***Horeyseck*** and ***Jänig*** 1974c, ***Jänig*** and ***Spilok*** 1978; SM: ***Jänig*** and ***Spilok*** 1978.
For 2. CVC, SM: ***Jänig*** and ***Kümmel*** 1981.
For 3. MVC: ***Horeyseck*** and ***Jänig*** 1974c; CVC, SM: ***Horeyseck*** and ***Jänig*** 1974c, ***Jänig*** and ***Kümmel*** 1981; ***Jänig*** and ***Spilok*** 1978.
For 4. ***Kümmel*** 1983

experimental conditions. Very occasionally, the resting activity of all MVC neurones is close to zero in a cat with a normal BP of about 100–110 mmHg. During recordings from single neurones, the rate of resting activity remains stable for hours. At present, it is unknown whether or not all adrenergic postganglionic neurones which innervate blood vessels of skeletal muscle have resting activity.

3.1.2 Stimulation of Arterial Baroreceptors

One of the most powerful inputs to the MVC system derives from arterial baroreceptors of the carotid sinus and aortic arch. This afferent influence is reflected in a pulsatile modulation of the activity in MVC neurones (cardiac rhythmicity of the activity; see post-R wave histogram of the activity in an MVC neurone in Fig. 3A). Pulsatile excitation of arterial baroreceptors leads to phasic inhibition of vasoconstrictor activity. Cutting the buffer nerves, or separating an open carotid blind sac with one

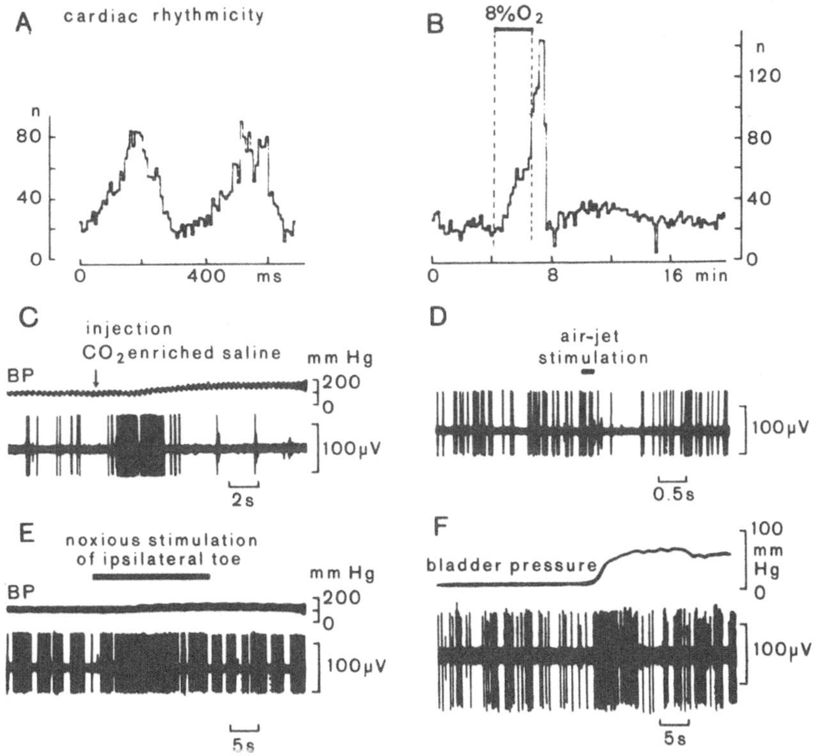

Fig. 3A–F. Discharge pattern in MVC neurones in brain-intact cats. **A–E** Same single postganglionic MVC neurone projecting to a peroneal muscle; **F** different postganglionic MVC neurone. **A** Post-R wave histogram of the MVC activity. Histogram obtained from 2000 trials (bin width 8 ms). **B** Stimulation of arterial chemoreceptors by making the animal breathe a gas mixture of 8% O_2 in N_2 for 3 min 30 s (bin width 10 s). **C** Stimulation of arterial chemoreceptors by a bolus injection of CO_2-enriched saline solution (0.8 ml) through a catheter in the lingual artery in the left carotid artery. **D** Short-lasting stimulation of hair follicle receptors on the trunk of the animal by air jets (10 trials superimposed). **E** Mechanical stimulation of nociceptors of a toe of the ipsilateral hindpaw. **F** Isovolumetric contraction of the urinary bladder. *n*, number of impulses per bin. (**A–C**, *Blumberg* et al. 1980; **D**, **E**, *Blumberg* and *Jänig*, unpublished observations; **F**, *Hilbers*, unpublished observation)

intact carotid sinus nerve (all other buffer nerves cut) from the arterial system, completely abolishes the cardiac rhythmicity of activity in MVC neurones. The post-R wave histograms under these surgically or functionally deafferentated conditions are flat and not higher than the peaks of activity in post-R wave histograms with open blind sac or with intact buffer nerves (*Gregor* et al. 1977; *Blumberg* et al. 1980), whether the mean BP (before deafferentation) was in the normal range of 100–120 mmHg or higher (up to 180 mmHg). Thus, the main inhibitory effect of arterial baroreceptor activity on the MVC system is phasic. This is also seen in the reactions of MVC neurones to supramaximal increase of static pressure in a carotid blind sac which was isolated from the arterial system (only one carotid sinus nerve to the blind sac was intact, all other buffer nerves were cut): the MVC activity is completely depressed during the first 10 s of the pressure step; thereafter, it recovers to values which are only some 10%–30% lower than before application of pressure (see also *Tedeschi* et al. 1971).

The quantitative analysis of the inhibitory influence of the arterial baroreceptors on the MVC activity in the cat leads to the following results: when one carotid sinus nerve to a carotid blind sac is intact, and the other carotid sinus nerve and both vagoaortic nerves are cut, the cardiac rhythmicity of the MVC activity is 90% ± 6% (mean ± 1 SE, n = 18). When the pulsatile excitation of the baroreceptor afferents is abolished by closing the carotid blind sac, with subsequent decrease of the BP in the blind sac to 60–100 mmHg, the MVC activity increases by 63% ± 6% (n = 18; *Blumberg* et al. 1980). When one or both carotid sinus nerves and both vagoaortic nerves are intact, the cardiac rhythmicity of the MVC activity is 131% ± 12% (mean ± 1 SE, n = 33; *Gregor* et al. 1977). The MVC activity increases by about 100% or somewhat less after cutting all buffer nerves.

Our results are in contrast to those of *Kendrick* et al. (1972), who claimed that complete unloading of baroreceptors leads to an increase of MVC activity to about 8 imp/s (≥ 400%). These authors used indirect methods and measured blood flow through skeletal muscle.

3.1.3 Stimulation of Arterial Chemoreceptors

Stimulation of arterial chemoreceptors by ventilating the animal with a gas mixture of 8% O_2 in N_2 (the respiration of the animal was "clamped" to a rate of 20/min), or by injecting a small bolus of CO_2-enriched saline solution through the lingual artery close to the glomus caroticum, always leads to an excitation of MVC neurones accompanied by an increase in BP (Fig. 3B, C; *Gregor* and *Jänig* 1977; *Blumberg* et al. 1980). During the arterial hypoxia (decrease of Pa_{O_2} to about 30 mmHg after the 2nd min from the start of ventilation with the hypoxic gas mixture; see *Weissman* et al. 1976), the activity in the MVC neurones increased by about two to eight times in the 2nd min (334% ± 53%; mean ± 1 SEM; n = 19; see *Blumberg* and *Jänig* 1984). This result is fully consistent with experiments reported in the literature in which changes of effector responses in skeletal muscle have been recorded, using various methods, during stimulation of arterial chemoreceptors in cats (*Weissman* et al. 1976; *Little* and *Öberg* 1975; *Wennergren* 1975), in dogs (*Daly* and *Scott* 1962; *Calvelo* et al. 1970; *Pelletier* 1972) and in rabbits (*Korner* 1971, 1979; *Korner* and *Uther* 1975).

Decerebration of the animals (leaving medulla oblongata and pons intact) does not change the responses of the MVC neurones to arterial hypoxia considerably (*Gregor* and *Jänig* 1977). These results are at variance with those of *Korner* and co-workers, who found that hypoxia, in pontine unanaesthetized rabbits, had almost no neural effect on the vascular bed of skeletal muscle (*Korner* 1971, 1979; *Korner* and *Uther* 1975). Also, arterial hypercapnia (ventilation of the animals with a gas mixture of 21% O_2 and 7.5% CO_2 in N_2) is followed by a slow increase of activity in MVC neurones (*Gregor* and *Jänig* 1977). We do not know whether this increase is due to stimulation of peripheral arterial chemoreceptors, central chemoreceptors (*Schlaefke* 1981) or both.

3.1.4 Stimulation of Cutaneous Receptors

Mechanical and thermal stimulation of cutaneous nociceptors induces reflex excitation in MVC neurones (Fig. 3E). This excitation is particularly obvious in a preparation in which aortic and carotid sinus nerves have been cut (*Horeyseck* and *Jänig* 1974b; *Jänig* 1975). In animals with intact buffer nerves, most MVC neurones are only initially excited during noxious stimulation of the skin since the increase of BP leads to excitation of arterial baroreceptors and this, in turn, to inhibition of MVC activity.

Stimulation of afferents from hair follicle receptors produces short-lasting depression of MVC activity (Fig. 3D). Stimulation of any other type of cutaneous receptors (Pacinian corpuscles, slowly adapting receptors, thermoreceptors) has no obvious effects on MVC neurones (*Horeyseck* and *Jänig* 1974a, *Jänig* 1975).

Both types of reflexes – that upon noxious cutaneous stimuli and upon stimulation of hair follicle receptors – can be elicited in MVC neurones from all over the body surface; they have no spatial organization, though the nociceptive reflex is most powerful when the toes of the paws and the tail are stimulated (*Horeyseck* and *Jänig* 1974a,b).

3.1.5 Stimulation of Visceral Mechanoreceptors

MVC neurones are generally excited by stimulation of afferents from the urinary bladder, colon and sometimes also the anus. The most efficient reflexes are elicited when the urinary bladder and colon contract isovolumetrically at intraluminal pressures of more than 20 mmHg (Fig. 3F). Each contraction is usually accompanied by an excitation of the MVC neurones. The degree of excitation varies from animal to animal. In some animals the level of MVC activity gradually increases during bladder and colon contractions without being related to the individual contractions, particularly when the contractions follow in rapid succession. Passive distensions of the organs, caused by an increase of intraluminal pressure of up to 100 mmHg, were much less effective than isovolumetric contractions in inducing reflex excitations of the MVC neurones. Important for these reflexes are the afferents through the pelvic nerve since cutting the hypogastric nerves leaves the reflexes almost unaffected (*Blumberg* et al. 1983).

3.1.6 Respiratory Modulation of Activity in MVC Neurones

One of the most characteristic features of the discharges in MVC neurones in our standardized experimental conditions (artificial ventilation at 20/s, tidal volume adjusted for an end-expiratory CO_2 of about 4%, both vagoaortic nerves and one carotid sinus nerve intact) is the respiratory rhythmicity. This modulation of the MVC activity is complex. The most important contributing factor is the central coupling between respiratory and "cardiovascular" neurones in the lower brainstem. Other factors are the influences from arterial baroreceptors in the carotid sinus and aortic nerves and from lung afferents in the vagal nerve on respiratory and cardiovascular neurones. The activity in these afferents changes with the artificial ventilation: lung afferents are activated during inspiration; arterial baroreceptor afferents are activated during rhythmic increase of BP, which is produced by the positive pressure ventilation (*Jänig* et al. 1980).

When the vagal (including aortic) or both vagoaortic and carotid sinus nerves are cut (thus abolishing the influence of activity in pulmonary stretch afferents on respiratory and cardiovascular neurones and, in the second case, the ventilation-related activation of baroreceptor afferents), the cycles of artificial ventilation and phrenic nerve discharges (which indicate inspiration) run independently. Under these conditions, the activity in MVC neurones increases during the phrenic nerve discharges (Fig. 4A). After the postganglionic discharge, activity is reduced to a minimum; then a second, slight reduction of activity may occur immediately prior to the peak of activity (see arrows in Fig. 4A). The postganglionic activity during the phrenic nerve discharge is enhanced by hypercapnia and by hypoventilation (decrease of tidal volume with consequent increase of P_{CO_2}) and probably corresponds to the excitation of Rα-inspiratory neurones. The depression following the postganglionic excitation may correspond to inhibition in Rα-inspiratory neurones produced by postinspiratory neurones; the slight depression preceding the postganglionic discharge may correspond to the initial inhibition of the Rα-inspiratory neurone discharge (see *Richter* 1982; *Richter* and *Ballantyne* 1983).

When vagal and buffer nerves are intact, the respiratory discharge pattern in MVC neurones is more complex. The activity profile usually has two peaks: one preceding the phrenic nerve discharge (*1* in Fig. 4B) and one following it (*2* in Fig. 4B). Peak 1 activity may be produced by the

Fig. 4A,B. Respiratory modulation of the activity in a strand with two to three postganglionic MVC axons which supplied the peroneal muscles. *Lower histograms,* activity in phrenic nerve recorded simultaneously. All histograms were obtained from 90 trials at a bin width of 64 ms. The end-expiratory CO_2 was 4 vol%. *Arrows* indicate short-lasting periods of depression. Note the two peaks of MVC activity when the vagal nerves are intact in **B**. *n*, number of impulses per bin. (Modified from *Jänig* et al. 1980)

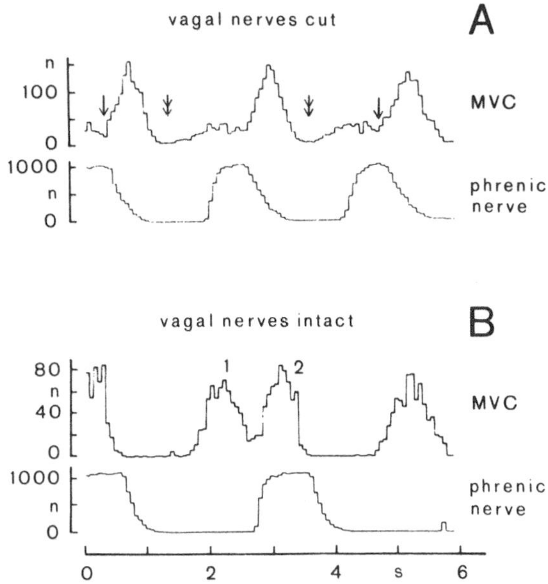

unloading of arterial baroreceptor afferents (produced by a decrease in BP induced by the positive pressure ventilation); peak 2 activity occurs during the phrenic nerve discharge and therefore during the discharge of Rα-inspiratory neurones. Peak 1 activity is not affected by hypercapnia but may increase during hyperventilation (increase of tidal volume at a respiratory rate of 20/min with subsequent increase in respiratory BP fluctuations). Peak 2 activity increases during hypercapnia and hypoventilation (decrease of tidal volume) and decreases during hyperventilation (*Jänig* et al. 1980). The gap of activity between both peaks may occur during the inhibition of Rα-inspiratory neurones at the beginning of the inspiratory phase (*Richter* 1982; *Richter* and *Ballantyne* 1983).

Experiments in which the inspiration of the artificial respiration cycle is coupled to the phrenic nerve discharge are necessary in order to evaluate the influence of excitation of afferents from pulmonary stretch receptors on the activity profile in MVC neurones. An important question concerns the biological meaning of the respiratory (and also cardiac! see Sect. 3.1.2) grouping of impulses in postganglionic MVC neurones for the regulation of vascular smooth muscle of resistance vessels and therefore for regulation of blood flow through skeletal muscle. At present we have no consistent ideas; we are, however, reluctant to believe that the rhythmic grouping of postganglionic discharges is biologically meaningless. Experiments on cats in which the sympathetic supply to skeletal muscle was stimulated electrically suggest that the most effective control of resistance and capacitance vessels may be attained by varying the intervals of burst and the intraburst frequency (*Andersson* 1983).

3.2 Vasoconstrictor Neurones Supplying Skin (CVC)

3.2.1 Ongoing Activity

The rate of ongoing activity in CVC neurones usually varies between about 0.1 and 2 imp/s and is rarely higher than 2 imp/s. It is unknown whether all CVC neurones or only a fraction of them have ongoing activity in the anaesthetized cat. The percentage of silent postganglionic neurones supplying the hairy skin of the cat hindlimb is probably higher than the 16% reported by *Grosse* and *Jänig* (1976). It might well be that both the rate of ongoing activity in individual CVC neurones and the fraction of CVC neurones with ongoing activity are similar in anaesthetized and unanaesthetized animals: in many experiments the temperature on the surface of the paw's skin was 24°–26°C at ambient temperatures of about 22°–23°C (body core temperature of the animals, 37.5°–38.5°C). The surface temperature may increase to 34°–36°C when the activity in CVC neurones is abolished (e.g. by cutting the LST or by hypothalamic warming or by anaesthetics, such as Althesin or halothane applied systematically; see Sect. 8).

The ongoing activity in CVC neurones has its origin, at least in part, in the hypothalamus: decerebrating a cat reduces ongoing activity in many CVC neurones (*Gregor* and *Jänig,* unpublished observations; *Gregor* and *Jänig* 1977). Mechanical damage of small regions in the medioanterior hypothalamus, with a thermoprobe, occasionally leads to a complete or partial decrease of activity in CVC neurones with a corresponding increase of skin temperature and a decrease of body core temperature (*Grewe, Jänig* and *Kümmel,* unpublished observations).

3.2.2 Stimulation of Arterial Baroreceptors

Stimulation of arterial baroreceptors has a weak or no inhibitory influence on most CVC neurones. This is suggested by little or no cardiac rhythmicity in the activity of most CVC neurones (Fig. 5A) and a weak transient or no depression of activity in CVC neurones when the arterial baroreceptors from an isolated carotid blind sac are stimulated by an increase of pressure (*Gregor* et al. 1977; *Blumberg* et al. 1980). These results are consistent with other investigations, conducted on carnivores, showing that blood flow and resistance to blood flow through skin does not change much during stimulation of arterial baroreceptors (*Brender* and *Webb-Peploe* 1969; *Folkow* and *Neil* 1971; *Kendrick* et al. 1972; *Ninomiya* et al. 1973; *Ninomiya* and *Irisawa* 1975; *Wennergren* 1975; *Hainsworth* and *Linden* 1979; *Karim* et al. 1980).

A minority of CVC neurones have high cardiac rhythmicity of their activity and are therefore under dominant control of arterial barorecep-

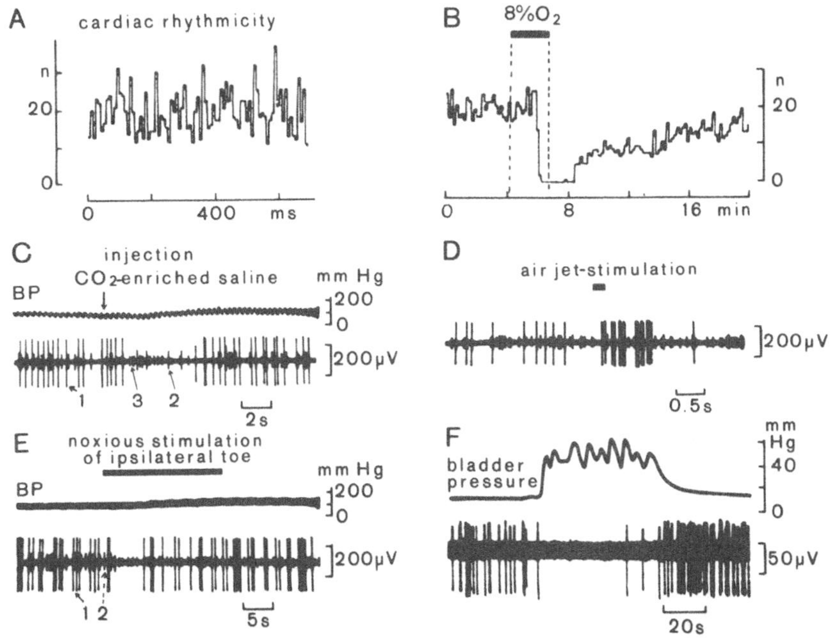

Fig. 5A–F. Discharge pattern in CVC neurones supplying hairy skin in brain-intact cats. **A–E** results obtained from a bundle with three postganglionic fibres (*1, 2, 3;* signal from fibre *3* only clearly seen in **C**); in **A** and **B**, only activity of CVC neurone *1* analyzed (same experiment as in Fig. 3A–E). **F** different postganglionic CVC neurone. **A** Post-R wave histogram of the activity in CVC neurone *1*. Histogram obtained from 1000 trials (bin width 8 ms). **B** Stimulation of arterial chemoreceptors by ventilating the animal with a gas mixture of 8% O_2 in N_2 for 2 min 30 s (bin width 10 s). **C–E** Same experiment as Fig. 3C–E. Note that neurone *3* was activated by stimulation of arterial chemoreceptors in **C** and that the depression of activity in neurone *2* outlasted the record in **E**. **F** Isovolumetric contraction of the urinary bladder. *n*, number of impulses per bin

tors (*Blumberg* et al. 1980). Furthermore, the degree of baroreceptor control is smaller for CVC neurones supplying hairless skin than for those supplying hairy skin (*Gregor* et al. 1977). This may mean that shunt and capacitance vessels of skin are innervated by CVC neurones which are under no or only minor control of arterial baroreceptors, whereas cutaneous "nutritional" vessels may be under dominant control (*Kendrick* et al. 1972; *Little* and *Öberg* 1975).

For the following reasons, it is doubtful whether the reflex control of CVC neurones by arterial baroreceptors, or its absence, can be integrated into a simple model:

1. Many CVC neurones show a short transient depression of activity as a reaction to an increase of pressure in an isolated carotid blind sac; this is followed by an overshoot of activity after the pressure increase. This *normally* does not happen in MVC neurones (*Blumberg* et al. 1980).

2. Occasional postganglionic neurones supplying skin respond with weak, delayed activation to stimulation of arterial baroreceptors by pressure steps in an isolated blind sac (*Blumberg* et al. 1980).

3. In cats with chronic skin nerve lesions on the hindlimb (cut and ligated superficial peroneal nerve; central stump of the superficial peroneal nerve sutured to the peripheral stump of a muscle nerve), the activity in CVC neurones projecting into the lesioned nerve, and to adjacent skin, may exhibit a higher degree of cardiac rhythmicity than obtained in animals without nerve lesions (*Blumberg* and *Jänig* 1983b, 1984).

These phenomena are poorly understood and indicate that the central organization of the CVC system in the neuraxis is complex with respect to the input from the baroreceptors.

3.2.3 Stimulation of Arterial Chemoreceptors

The activity of most CVC neurones supplying the hairy skin and of all CVC neurones supplying the hairless skin of the cat paw is depressed by systemic hypoxia (Fig. 5B). Direct stimulation of arterial chemoreceptors, by i.a. injection of a small bolus of CO_2-enriched Tyrode's solution close to a glomus caroticum, also leads to a short-lasting depression of activity in most CVC neurones (Fig. 5C). The responses to systemic hypoxia are abolished or reversed to small excitations after cutting the carotid sinus and aortic nerves (*Gregor* and *Jänig* 1977; *Jänig* and *Kümmel* 1977, 1981; *Blumberg* et al. 1980). These results are fully consistent with other investigations in which postganglionic activity in nerve twigs supplying the rabbit ear (*Iriki* et al 1971a,b), or thermal conductivity of skin, cutaneous venous outflow and perfusion pressure of superficial veins, were measured (*Korner* 1971, 1979; *Korner* and *Uther* 1975; *Mancia* 1975; *Pelletier* 1972; *Pelletier* and *Shepherd* 1972, 1975; *Weissman* et al. 1976).

Little and *Öberg* (1975), however, reported that, in the cat, stimulation of arterial chemoreceptors probably does not influence CVC neurones to shunt vessels (arteriovenous anastomosis) but does lead to activation of CVC neurones supplying nutritional (resistance) vessels. Also, *Calvelo* et al. (1970) reported that, in the dog, stimulation of arterial chemoreceptors leads to adrenergic reflex constriction of veins in the hindpaw, whereas prevenous vessels in the paw are dilated by activation of non-adrenergic vasodilator neurones during this type of stimulus (see Sect. 3.6). These results on chemoreceptors reflexes in the adrenergic vasoconstrictor supply to the skin are apparently irreconcilable with the other investigations cited above and with our own investigations, in particular, the investigation of the vasoconstrictor supply to the hairless skin of the paw (*Jänig* and *Kümmel* 1977, 1981). Perhaps different sections of the cutaneous vascular bed (shunt, capacitance, nutritional vessels) are regulated by different types of vasoconstrictor neurones (*Little* and *Öberg* 1975; see *Blumberg* et al. 1980): first, shunt and cutaneous capacitance vessels by vasoconstrictor neurones, which are inhibited during stimulation of arterial chemoreceptors (and which are not or only weakly controlled by arterial baroreceptors; see Sect. 3.2.2); second, cutaneous nutritional (resistance) vessels by vasoconstrictor neurones, which are excited during stimulation of arterial chemoreceptors (and which are under strong control of arterial baroreceptors; see Sect. 3.2.2).

Alternatively, it may be hypothesized that the CVC system is under inhibitory, as well as excitatory, control of the arterial chemoreceptors and that the balance between these influences determines the response seen in the postganglionic neurones. For example, during our standardized experimental conditions (see Sect. 2.2), the excitatory influence from the arterial chemoreceptors on the CVC system may usually be completely masked in brain-intact animals. CVC neurones are only occasionally excited during stimulation of chemoreceptors or exhibit a dual response (excitation followed by inhibition; *Gregor* and *Jänig* 1977; *Blumberg* et al. 1980). After decerebration, CVC neurones are excited by stimulation of arterial chemoreceptors (*Gregor* and *Jänig* 1977; *Iriki* and *Kozawa* 1976).

3.2.4 Stimulation of Cutaneous Receptors

Innoxious Stimuli. Air jets, which stimulate group II and III afferents from hair follicles, lead to short-lasting reflex excitation of most CVC neurones. This is usually followed by depression of resting activity of variable duration (Fig. 5D). Some CVC neurones are slightly depressed in their activity without preceding excitation and some are not affected at all. Others are repetitively excited at intervals of 200–500 ms. This reflex is easily overlooked since it is short lasting and can usually only be detected when several trials are superimposed (*Horeyseck* and *Jänig* 1974a; *Jänig* 1975; *Grosse* and *Jänig* 1976). It can be obtained from all over the body surface (*Horeyseck* and *Jänig* 1974a). It may be part of an abortive arousal reaction which is largely suppressed by the general anaesthesia of the animals. It requires the integrity of the upper mesencephalon and hypothalamus (*Jänig* 1975).

A second reflex can be elicited in some 30% of CVC neurones supplying the hairless skin of the cat paw. If Pacinian corpuscles in the paw are stimulated by vibration of the paw (produced by tapping against the paw holder), activity in the CVC neurones is depressed (*Jänig* and *Kümmel* 1977, 1981). This reflex has either been overlooked previously or is absent in CVC neurones supplying hairy skin (*Horeyseck* and *Jänig* 1974a). Other innoxious cutaneous stimuli, such as cold and warmth applied to the hindlimbs (20°–43°C), thus exciting cold and warm receptors, and constant

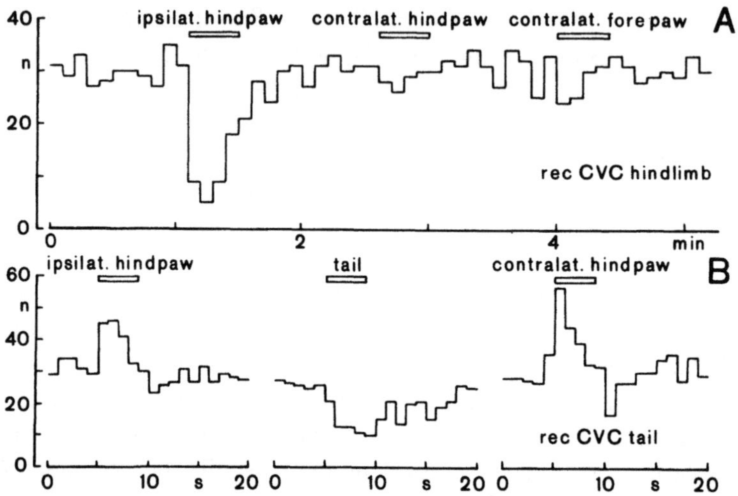

Fig. 6A,B. Reactions of CVC neurones to mechanical noxious stimulation of skin. **A** Recording of activity from three postganglionic CVC neurones supplying hairy skin of the hindlimb (SU; bin width 6 s; ordinate scale impulses per bin). **B** Recording from a postganglionic CVC neurone supplying hairy skin of the tail (bin width 1 s; 10 trials superimposed; ordinate scale number of impulses per bin). (**A** modified from *Jänig* 1975; **B** modified from *Grosse* and *Jänig* 1976)

force stimuli exciting slowly adapting mechanoreceptors, are largely without effect on activity in CVC neurones (*Horeyseck* and *Jänig* 1974a,b).

Noxious Stimulation of the Skin. In most CVC neurones, noxious stimulation of the skin elicits depression of resting activity (Fig. 5E). This unique reflex was originally described in rabbits by *Lovén* (1866). *Lovén* wrote that, after electrically stimulating the central stumps of cut skin nerves to the ear and to the hindfoot, he observed dilation of adjacent cutaneous arteries which are innervated by other nerves (Lovén reflex). He concluded, in his remarkable paper, that the dilation is a reflex and mediated by nerve fibres. The Lovén reflex is certainly not an axon reflex which is mediated by antidromic excitation of branching afferent fibres of which one cut branch was stimulated electrically. Such types of high-branching afferents could not be shown to exist in skin nerves of the cat hindlimb (*Blumberg* and *Jänig*, unpublished observations).

The depression of resting activity induced by noxious stimulation of skin has several characteristics (*Horeyseck* and *Jänig* 1974b; *Jänig* 1975; *Grosse* and *Jänig* 1976; *Jänig* and *Kümmel* 1977, 1981): it is most powerful when nociceptors of the distal parts of the extremities (toes!) and tail are stimulated mechanically or by radiant heat. Frequently, it considerably outlasts the end of the stimulus (see unit 2 in Fig. 5E). The reflex has a spatial organization (Fig. 6A,B): it is largest when the skin of the paw,

Table 3. Incidence of reflexes in CVC neurones induced by mechanical noxious stimulation of skin (toes or ipsilateral hindpaw)

Series[a]	↓	↑	∅	*n*
1	56%	30%	14%	50
2	87%	13%	0	46
3	100%	0	0	39

↓, depression; ↑, excitation; ∅, no effect; *n*, number of postganglionic axon bundles analyzed

[a] Series 1, CVC neurones supplying hairy skin (SP and SU; *Horeyseck* and *Jänig* 1974b); series 2, CVC neurones supplying hairy and hairless skin (SP, SU, MP; *Hilbers* and *Jänig,* unpublished observations); series 3, CVC neurones supplying hairy skin of tail (*Grosse* and *Jänig* 1976) and hairless skin of hindpaw (*Jänig* and *Kümmel* 1977)

which is supplied by the CVC neurones, is stimulated and smaller or absent when the skin of the contralateral paw and of the forepaws is stimulated. Some CVC neurones supplying the tail were depressed during noxious stimulation of the tail's skin and excited during noxious stimulation of the toes of the ipsi- and contralateral hindpaws (Fig. 6B). The incidence of this nociceptive reflex in CVC neurones is shown in Table 3 for three experimental series. In the first series by *Horeyseck* and *Jänig* (1974b), carotid sinus nerves and vagoaortic nerves were cut in order to abolish the influence of activity in arterial baroreceptors. In the second and third series, both vagoaortic nerves and one carotid sinus nerve were left intact. At present, it is unclear why the results differ between series 1 and the other two series. The control of the animals in the first series was probably not as precise as in the later two.

3.2.5 Stimulation of Visceral Afferents

Activity in CVC neurones is depressed during stimulation of afferent fibres from the urinary bladder, colon and anus. Isovolumetric contractions of the urinary bladder and colon are most effective for producing these inhibitory reflexes; passive distensions of the organs and mechanical stimulation of anal skin are much less effective. Usually each contraction of the organs is accompanied by a depression of activity in CVC neurones (Fig. 5F). These reflexes are quantitatively very variable from animal to animal. Sometimes the activity in the CVC neurones decreases gradually when the organ is filled slowly over periods of 10 min or longer without inducing contractions. It must be assumed that – in these experiments – the slowly increasing activity in visceral afferents induces the slow decrease of activity in CVC neurones. Important for the inhibitory viscerosympathetic reflexes in CVC neurones is the integrity of pelvic visceral afferent fibres

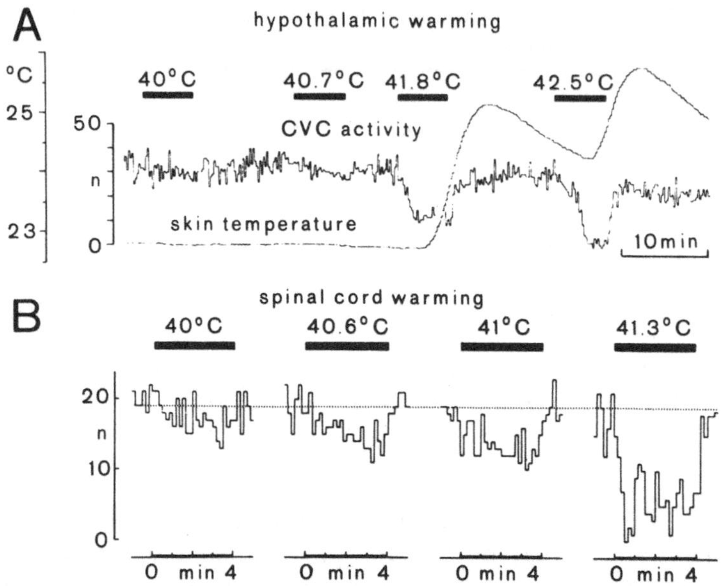

Fig. 7A,B. Reaction of CVC neurones to warming of the hypothalamus and spinal cord. **A** Warming of the hypothalamus by a water-perfused thermode, which was positioned medially into the anterior hypothalamus. Temperatures about 1 mm lateral to thermode. The activity was recorded from a bundle with two to three postganglionic CVC axons which innervated the hairless skin of the central pad of the left paw (bin width 10 s). The skin temperature was measured on the surface of the central pad. **B** Warming of the spinal cord by a water-perfused, U-shaped tubing positioned epidurally dorsal to the spinal cord. The temperatures on the dorsal surface of the spinal cord are indicated (values taken from *Simon* and *Iriki* 1971). Activity recorded from single CVC neurone supplying hairy skin (bin width 10 s). *n,* number of impulses per bin. (**A**, *Grewe, Jänig, Kümmel* and *Xu*, unpublished observation; **B**, modified from *Gregor* et al. 1976)

to the sacral spinal cord. Cutting the hypogastric nerves has no obvious effect on these reflexes (*Hilbers* and *Jänig,* unpublished observations). This does not mean that visceral afferent fibres in the hypogastric nerves, which supply pelvic organs, do not participate in these reflexes (*Blumberg* et al. 1983).

3.2.6 Thermal Stimulation of Spinal Cord and Hypothalamus

It is generally accepted that CVC neurones, at least those supplying arterovenous anastomoses and cutaneous capacitance vessels (*Mellander* and *Johansson* 1968), are involved in thermoregulation. It is furthermore accepted that the anterior hypothalamus and the spinal cord have temperature-sensitive and thermoregulatory neurones which, when excited during temperature stimuli, can elicit thermoregulatory responses in the periphery (*Thauer* 1970; *Simon* 1974; *Hensel* 1981). Warm stimuli applied to both CVC structures lead to depression of resting activity in CVC neurones supplying hairy and hairless skin. This depression is graded when the stimulus strength is graded (Fig. 7). Warming the hypothalamus and spinal cord simultaneously leads to responses which are larger than the algebraic sum of the individual responses. Cold stimuli applied to the spinal cord and hypothalamus are much less effective than warm stimuli and usually lead to small transient increases of activity in CVC neurones. The depression of activity in CVC neurones during warming of the spinal cord and hypothalamus is selective: MVC neurones are either not, or only indirectly, affected due to a decrease of systemic BP or due to activation of respiratory neurones being coupled with vasomotor neurones in the lower brainstem.

The decrease of resting activity in CVC neurones during central warm stimuli is accompanied by an increase of temperature at the skin surface (Fig. 7A). The skin temperature increases only when the resting activity in CVC neurones decreases by more than 40%. At present it cannot be explained why the skin temperature does not change when the activity in the CVC neurones decreases by up to 40%. It is probably not due to the fact that vasodilator neurones are switched on during critical hypothalamic warm stimuli. We did not find postganglionic neurones projecting into skin nerves of the hindlimb which could be activated by hypothalamic warm stimuli (see Sect. 3.6). It may be due to the relatively low environmental temperature (about 22°C), which may influence the neuro-effector transmission or the contractility of the vascular smooth muscle (*Vanhoutte* 1980); the response curve of the vascular smooth muscle to graded neuronal input may be shifted to the left in a cold environment so that a decrease of activity in vasoconstrictor neurones by 40%–50% does not lead to an increase of blood flow. We are sceptical about this explanation since we obtained the same results at skin surface temperatures of 22° to about 30°C (*Grewe* et al. 1982).

3.2.7 Respiratory Modulation of Activity in CVC Neurones

The activity in CVC neurones exhibits some respiratory modulation (operational definition, see Sect. 3.1). The degree of this respiratory modulation is smaller than that of the respiratory modulation of activity in MVC neurones (Sect. 3.1). Furthermore, it is smaller in CVC neurones supplying hairless skin than in CVC neurones supplying hairy skin (*Gregor* et al. 1977). Some CVC neurones, however, have respiratory modulation of their activity which is as pronounced as in MVC neurones. The respiratory modulation of CVC neurones rarely shows the two peaks in their activity profile, which are typical for MVC neurones (see Fig. 4B). Hypercapnia (ventilation of the animal with a gas mixture of 7.5% CO_2 in

21% O_2 and 71.5% N_2) and hypoventilation (decrease of tidal volume without changing ventilation frequency) somewhat increase the activity during the phrenic nerve discharges (i.e. during inspiration) in animals in which vagoaortic and carotid sinus nerves are intact, as well as in animals in which these nerves are cut (*Jänig* et al. 1980).

3.3 Sudomotor (SM) Neurones

It is difficult to isolate and recognize postganglionic axons with putative SM function in the plantar nerves. First, the SM system is very sensitive to any type of general anaesthesia (*Jänig* and *Räth* 1977, 1980). Second, sweat glands are restricted to the hairless skin of the paws of the cat feet and, therefore, the proportion of axons with SM function in the plantar nerves is probably low with respect to the total number of postganglionic CVC axons. These technical problems were overcome by selecting fascicles from the medial plantar nerve (MP) which innervate the central pad of the hindpaw and by recording the skin potential from the surface of the hairless skin of the central pad (*Jänig* and *Kümmel* 1977).

3.3.1 Ongoing Activity in SM Neurones and Its Correlation with Negative Transient Skin Potentials

Under the usual experimental conditions (see Sect. 2.1), postganglionic SM neurones display either very low ongoing activity of up to 0.4 imp/s (mean ± 1 SD:0.2 ± 0.1 imp/s, see *Jänig* and *Kümmel* 1977) or are silent. Many of the ongoing discharges are followed by transient negative skin potentials at stable latencies of 600–800 ms (Fig. 8A,B). The negative transient skin potentials are most likely produced by activation of the secretory sweat gland cells; they are abolished by atropine. The correlation between single impulses in individual SM units and the transient skin potentials (which is not always as clear-cut as in Fig. 8A,B) requires that many postganglionic SM neurones discharge in synchrony.

The low rate of ongoing activity or its absence in SM neurones is probably due to the chloralose anaesthesia. In some animals no resting activity at all (and no reflex activity) could be recorded in the SM system, and sometimes one had to wait for 1–3 h after surgery before resting and reflex activity appeared in the SM system. During this period, no obvious change of depth of anaesthesia seemed to occur, as judged by the width of the pupils. Usually the best results are obtained from animals which had almost completely miotic pupils and which exhibited a stable BP and heart rate.

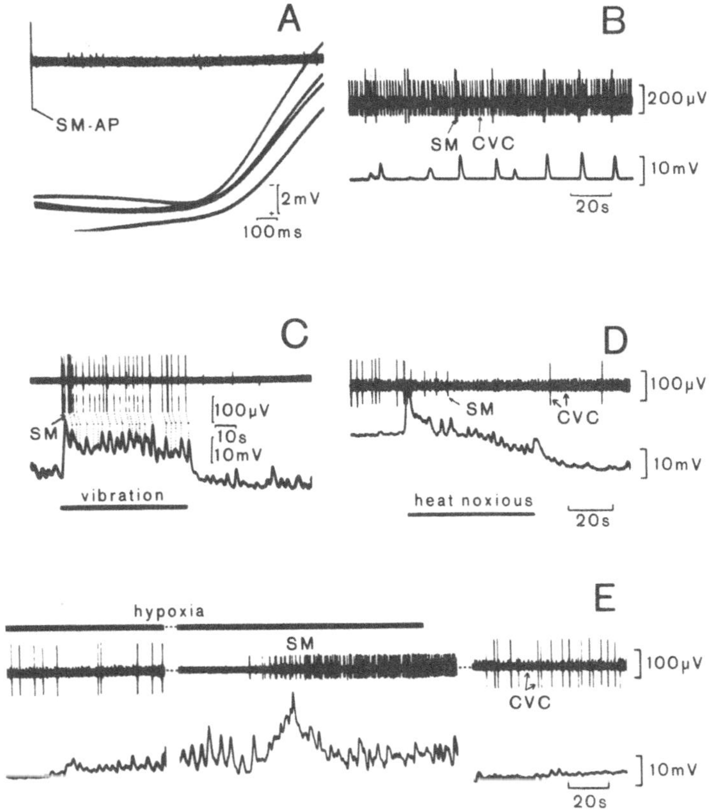

Fig. 8A–E. Discharge pattern in SM neurones in brain-intact cats. *Upper records,* post-ganglionic activity. *Lower records,* skin potential (negative up). **A, B** Ongoing activity; in **A** the oscilloscope trace was started with the action potential of an SM neurone (*SM-AP*); in **B** most discharges of an SM neurone are followed by transient skin potentials. The CVC activity was not correlated with the skin potential. **C** Reflex activation of a single SM neurone by a vibrational stimulus (stimulation of Pacinian corpuscles in the paws by tapping on the experimental frame). Note that most discharges of the SM neurone are followed by fast deflections of the skin potential. **D** Reaction of an SM neurone and two CVC neurones to noxious stimulation of a toe of the ipsilateral hind paw by radiant heat (55°C). **E** Reaction of an SM neurone and two CVC neurones to stimulation of arterial chemoreceptors by ventilating the animal with a gas mixture of 8% O_2 in N_2 (the ventilation started about 1 min before the record). Same preparation as in **D**. (**A, B** and **D**, modified from *Jänig* and *Kümmel* 1977; **C**, *Jänig* 1979; **E**, modified from *Jänig* and *Kümmel* 1981)

3.3.2 Stimulation of Cutaneous Receptors

A unique and powerful activation of SM neurones can be attained by stimulation of *Pacinian corpuscles* in the paws, particularly in the hind-paws (Fig. 8C). Any stimulus which excites these receptors may lead to activation of SM neurones, e.g. vibration, tapping on the hindpaw holder, air-jet stimuli applied to the skin area between the toes, sometimes even walking through the room or lightly tapping on the experimental frame (*Jänig* and *Räth* 1977; *Jänig* and *Kümmel* 1977). Individual deflections of the evoked skin potential may be time-locked to the preceding SM discharges (Fig. 8C). This type of activation of the SM neurones is not just part of an arousal reaction: air-jet stimuli applied tangentially to the hairs of limbs and trunk and loud, high-frequency auditory stimuli do *not* lead to activation of the SM neurones (*Jänig* and *Räth* 1977; see also Sect. 8.1).

The vibration reflex is largest (as judged from the size of the skin potential) in the SM neurones supplying the hindpaws, when the Pacinian corpuscles in the hindpaws are stimulated. It is small in the SM neurones supplying the forepaws (*Jänig* and *Räth* 1977). The discharge rates of SM neurones during vibration stimuli, which elicit transient skin potentials of 10–20 mV, are small and range from 0.2 to 2.1 imp/s (mean ± 1 SD: 0.6 ± 0.5 imp/s, $n = 21$; *Jänig* and *Kümmel* 1977).

The function of this "vibration reflex" in the SM neurones may be to keep the surface of the hairless skin of the paw soft and flexible. This may increase friction during contact and may create optimal conditions for somatosensory perception since each contact of the paws with surfaces during movements and manipulations excites the Pacinian corpuscles. A second cutaneous stimulus which leads to excitation of SM neurones is stimulation of *cutaneous nociceptors* produced by radiant heat or by pinching the skin. This reflex excitation of the SM neurones is smaller than the excitation induced by stimulation of Pacinian corpuscles (Fig. 8D). The reflexes in SM neurones supplying the hindpaws which are elicited by noxious stimulation of the skin exhibit no spatial organization (*Jänig* and *Räth* 1977; *Jänig* and *Kümmel* 1977).

Other stimuli applied to the skin which lead to excitation of hair follicle afferents (air jet stimuli), of slowly adapting receptors in the paws (constant force stimuli) and of cold and warm receptors (thermal stimuli) have no obvious effect on the activity in SM neurones.

3.3.3 Stimuli Applied to Receptors in the Interior of the Body

The activity in SM neurones displays no rhythmic changes with respect to the pulsatile excitation of *arterial baroreceptors*, indicating that the pulsatile excitation of these afferents has at least no phasic influence of SM neurones (*Jänig* and *Kümmel* 1977). *Horwitz* and *Kaufman* (1979) found, in some 40% of their experiments, that stimulation of arterial baroreceptors in isolated blind sacs led to some depression of spontaneous skin potentials measured on the surface of the hairless skin of the cat hindpaw.

Systemic hypoxia (ventilation of the animals with a gas mixture of 8% O_2 in N_2) and systemic hypercapnia (ventilation with a gas mixture of

7.5% CO_2 and 21% O_2 in N_2) lead to large activations of the SM neurones (*Jänig* and *Kümmel* 1977, 1981). These stimuli excite peripheral *chemoreceptors* and – as far as the hypercapnia is concerned – probably central chemoreceptors (*Schlaefke* 1981). The responses of SM neurones are always reciprocal to those in CVC neurones supplying the hindpaw (Fig. 8E).

The activity in SM neurones exhibits some *respiratory rhythmicity*. The rhythmicity is enhanced considerably by hypercapnea (*Jänig* and *Kümmel* 1977; *Jänig* et al. 1980). It is probably due to central coupling between respiratory neurones and the SM system. Better controlled experiments on animals with cut carotid sinus and/or vagodepressor nerves and with simultaneous recording of activities in SM neurones and in the phrenic nerve are necessary to quantitate the degree of central coupling between the respiratory and SM systems.

Stimulation of *afferents* from the urinary bladder and colon by isovolumetric contractions of the organs also leads to excitation of SM neurones (*Blumberg* et al. 1983). Each contraction of the organs is followed by transient negative skin potentials on the hindpaw.

Central warm and cold stimuli applied to the spinal cord and hypothalamus either do not influence the activity in SM neurones or lead to short-lasting excitations (*Grewe* et al. 1982). It is likely that the SM system of the cat is not involved in thermoregulation (but see *Wang* and *Chun* 1967).

3.4 Pilomotor (PM) Neurones

Erector pili muscles seem to exist in the hairy skin of the cat all over the body surface. However, these smooth muscles are well developed only in the hairy skin of the tail, back and dorsal part of the head (*Strickland* and *Calhoun* 1963), and only here can piloerection be observed in this animal during different behavioural states and during electrical stimulation of different sections of the sympathetic chain (*Langley* and *Sherrington* 1891; *Langley* 1894; *Lewis* and *Marvin* 1927; *Bopp* 1954; *Hunsperger* 1965; *Leyhausen* 1979). Thus, only the erector pili muscles in the above-mentioned areas seem to be innervated by postganglionic neurones.

Nerves to the skin of the cat tail contain postganglionic axons with ongoing activity which behave like CVC neurones (see Sect. 3.2) and many unmyelinated silent postganglionic axons with high conduction velocity (see Sect. 7.3) (*Grosse* and *Jänig* 1976). These silent postganglionic neurones cannot be activated by any natural somatic stimulus, a few postganglionic silent neurones can be driven very weakly by strong electrical stimuli applied to a hindlimb nerve. Only asphyxia (termination of the artificial ventilation) leads to an activation of these neurones after 3–4 min. With the beginning of this activation, piloerection regularly occurs on the tail; when the discharge of these postganglionic neurones ceases, the piloerection also ceases. This pattern of activation of PM neurones is different from the pattern occurring in CVC neurones during asphyxia. Most CVC neurones are first inhibited and then excited; this excitation precedes that of the PM neurones (*Grosse* and *Jänig* 1976). It is unknown whether PM neurones are activated by a cold environment. It could well be that piloerection observed in cats in cold environments is produced by an increase of sensitivity of erector pili muscles to circulating catecholamines (*Hellmann* 1963).

It is likely that PM neurones are activated particularly during certain species-specific behaviour and that they are predominantly controlled by the hypothalamus and the limbic system and normally not influenced by stimuli applied to the body surface. In hypothalamic animals, however, noxious and innoxious somatic stimuli may lead to piloerection (*Bard* 1928; *Bard* and *Rioch* 1937; *Ectors* 1941; *Bard* and *Macht* 1958).

3.5 Vasodilator Neurones Supplying Skeletal Muscle (MVD)

The postganglionic sympathetic supply to skeletal muscle of the cat hindlimb contains non-adrenergic fibres which when *activated* induce dilation of arteriolar resistance vessels. This neuronally induced increase of blood flow can be abolished by atropine (but see *Bell* 1983); thus, it is probably produced by activation of postganglionic cholinergic vasodilator neurones (*Burn* 1938; *Uvnäs* 1954; but see *Bell* 1983). These neurones should have their somata in the sympathetic trunks and receive their synaptic input from preganglionic axons having their somata in the lumbar spinal cord (see *Sonnenschein* and *Weissman* 1978). Repetitive electrical stimulation in the hypothalamic defence area in anaesthetized cats also elicits atropine-sensitive vasodilation in skeletal muscles of hindlimbs in conjunction with other autonomic responses (e.g. increase of cardiac output, reduction of blood flow through the skin and intestine). It is believed that the cholinergically induced vasodilation in skeletal muscle is an integral part of the defence reaction which is organized at the hypothalamic level (*Eliasson* et al. 1951; *Abrahams* et al. 1960, 1964; *Uvnäs* 1954, 1960, 1967; *Lisander* 1970). Recent investigations on unanaesthetized cats demonstrated that "emotional stimuli" (confrontation of a cat with a dog or with another cat) can lead to cholinergic vasodilation in the skeletal muscle. This vasodilation consists of two components: a diffuse vasodilation occurring in all limbs and a vasodilation which is restricted to the skeletal muscle of that limb which in the next moment will strike at the target.

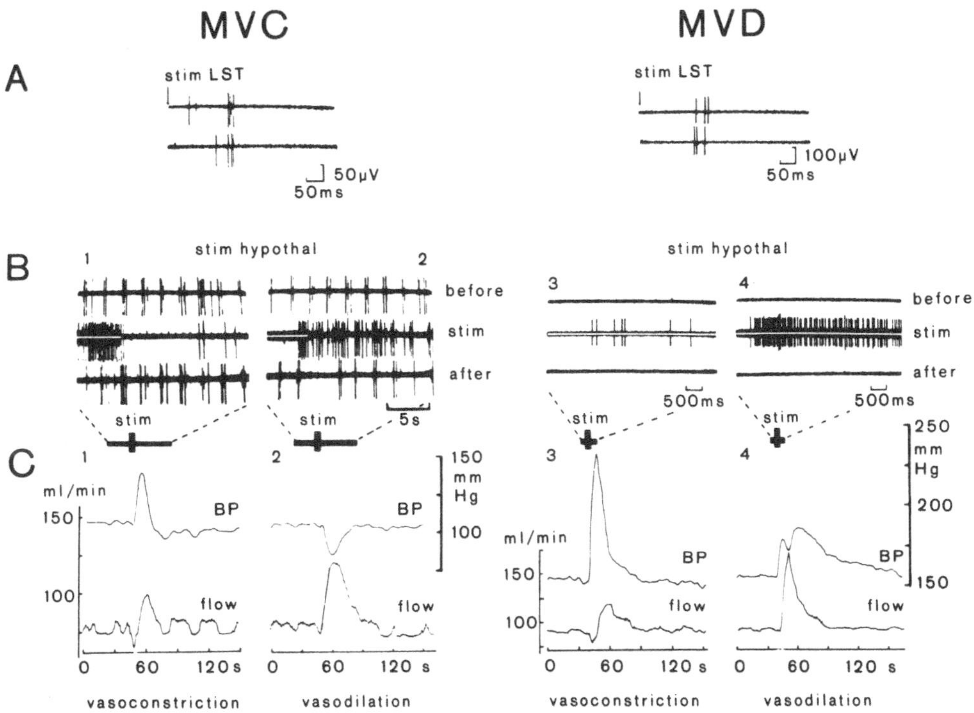

Fig. 9A–C. Reaction of an MVC and an MVD neurone to bipolar electrical stimulation of the hypothalamus inducing vasoconstriction and atropine-sensitive vasodilation in the hindlimbs, respectively. The neural activity was recorded from postganglionic axons supplying the gastrocnemius-soleus muscle. Blood flow was measured 1 cm proximal to the aortic bifurcation using an electromagnetic flow probe. **A** Stimulation of the preganglionic axons in the LST between ganglion L_4 and L_5 with single pulses (5 V, 0.2 ms). **B** Neural activity before, during and after bipolar stimulation of the hypothalamus (*white bars*). Stimulation parameters, 8–10 V (pulse duration 1 ms), 100 Hz for 5 s. Stimulation sites, *1* (vasoconstriction), between (frontal) F_{10} and F_{12}, (ventral) V_6, (lateral) L_2. *2* (vasodilation), between F_8 and F_{10}. V_6, L_2. *3* (vasoconstriction), F_{12}, V_5, L_2. *4* (vasodilation), F_8, V_5, L_2. **C** *BP* and blood flow (*flow*) before, during and after stimulation of the hypothalamus (*stim hypothal*). The duration of the stimulation is indicated by the *vertical bars;* the *horizontal bars* below **B** correspond to the duration of the specimen records in **B**. (Modified from *Horeyseck* et al. 1976)

This experiment implies that the cholinergically induced vasodilation in skeletal muscle requires very complex environmental (species-specific) stimuli and that the central neuronal machinery inducing it has some spatial organization (*Mancia* et al. 1972; *Ellison* and *Zanchetti* 1973; *Zanchetti* et al. 1972).

Electrophysiological investigations show that about 10% of the postganglionic neurones supplying the lateral gastrocnemial muscle are activated during atropine-sensitive vasodilation induced by electrical stimulation of the hypothalamic defence area (Fig. 9B4, C4). The activated neurones exhibit no ongoing activity (Fig. 9B3, 4), no reflex activity to any somatic stimuli, no reflex activation to chemoreceptor stimulation by asphyxia and no depression of the hypothalamically induced responses by stimulation of arterial baroreceptors. Electrical stimulation of a hypothalamic area which leads to vasoconstriction in skeletal muscle has either no effect on these silent postganglionic neurones or leads to a weak excitation of them (Fig. 9B3, C3) (*Horeyseck* et al. 1972, 1976; but see *Takeuchi* and *Manning* 1971, 1973). These functional properties are characteristic for MVD neurones. It was striking that the signals recorded from these putative postganglionic MVD axons are always larger than those recorded from postganglionic MVC axons. This indicates that MVD axons may conduct faster than MVC axons (*Horeyseck* et al. 1976; see Sect. 7.3).

In contrast, the reactions of MVC neurones to electrical stimulation of the hypothalamus are reciprocal to those of the MVD neurones: during hypothalamically induced vasoconstriction, MVC neurones are activated followed by a depression of ongoing activity (Fig. 9B1, C1); during hypothalamically induced atropine-sensitive vasodilation, MVC neurones are depressed or transiently depressed, and this depression is sometimes followed by enhanced activity (Fig. 9B2, C2).

3.6 Vasodilator Neurones Supplying Skin (CVD)

After abolishing the adrenergic vasoconstrictor effect on peripheral vessels by guanethidine, electrical stimulation of the LST with trains of pulses elicits an increase of temperature on the surface of the hairless skin of the hindpaw in adrenalectomized cats. This increase of temperature is due to dilation of vessels in the skin; it is resistant to atropine and blocked by hexamethonium at doses which do not or only marginally block the impulse transmission to postganglionic SM neurones (*Bell* et al. 1985). This may mean that the hairless skin of the cat paw is supplied by a third type of postganglionic neurone which, when activated, produces a vasodilation in the skin via a non-cholinergic, non-adrenergic mechanism. Alternatively, it may mean that this vasodilation is produced by vasoactive intestinal polypeptide (VIP), which coexists with acetylcholine in SM neurones and is released by SM axon terminals when activated (*Hökfelt* et al. 1980; *Lundberg* 1981). Evidence against the latter possibility may be that the vasodilation produced by activation of postganglionic SM neurones is probably restricted to the tissue around the sweat glands deep below the surface of the skin, that reflex activation of SM neurones by vibrational stimuli (see Sect. 3.3.2) does not lead to an increase of temperature on the surface of the hairless skin (*Jänig* and *Kümmel*, unpublished observation), and that hexamethonium in low doses does block the vasodilator response but not the SM response (*Bell* et al. 1985). Furthermore, the skin of the dog paw appears to be supplied by noncholinergic, nonadrenergic vasodilator neurones which may be involved in adaptive cardiovascular responses during stimulation of arterial chemoreceptors (*Calvelo* et al. 1970; *Abboud* et al. 1975, 1976) and/or in thermoregulatory responses (*Schönung* et al. 1972; *Peter* and *Riedel* 1982; *Bell* 1983). *Schönung* et al. (1972) showed that warming of the spinal cord leads to a sympathetically mediated increase of blood flow through the paw skin in dogs treated with reserpine. *Bell* (1983) believes that a candidate for a transmitter mediating the increase of blood in the paw skin of the dog is dopamine (for lit. see *Bell* 1983).

Using the thermoregulatory paradigm of *Schönung* et al. (1972), an attempt was made to search for postganglionic neurones in peripheral skin nerves of the cat hindlimb which can be activated by spinal cord warming. This would be in contrast to CVC neurones, which are inhibited (see Sect. 3.2.6). In the initial attempt, unmyelinated axons were found which were activated in a graded manner to warming of the spinal cord (temperature at the surface of the spinal cord about 41°–43.5°C; *Gregor* et al. 1976). In the second attempt, the results were confirmed in brain-intact and in chronic spinal cats, but it was not possible to block the responses with hexamethonium or to activate these responding neurones by stimulation of the sympathetic trunk (*Jänig* and *Kümmel* 1981). The warm-sensitive neurones did not respond to any other stimuli tested so far (stimulation of arterial chemoreceptors and cutaneous receptors; warm stimuli applied to the hypothalamus; but see *Peter* and *Riedel* 1982). Most of the units had no resting activity; three of eight unmyelinated warm sensitive units in chronic spinal animals had resting activity at a normal body core temperature of 38°C, which ceased upon spinal cord cooling. The maximal discharge rates during warming of the dorsal surface of the spinal cord to about 43°C for 3–4 min amounted to about 0.3 3.2 imp/s (mean 1.8 ± 0.9 imp/s, n = 9). The incidence of these types of warm-sensitive units appears to be low (*Gregor* et al. 1976; *Jänig* and *Kümmel* 1981).

Given the graded responses of these warm-sensitive units to warming of the spinal cord and the resting activity at normal body core temperature in some of these units, it is unlikely that an artefact has been produced. Experiments are needed in order to clarify whether these warm-sensitive units are afferent or whether they are postganglionic and perhaps driven via a non-nicotinic synaptic mechanism by preganglionic axons which pass through the lower lumbar ventral roots (*Graham* and *Lioy* 1973). Furthermore, it must be clarified whether a non-cholinergic, non-adrenergic active neurogenic vasodilation can be elicited in the skin of the cat paw by natural peripheral or central stimuli, as is the case for active vasodilation in skeletal muscle and in the tongue skin (*Pleschka* 1984).

3.7 Other Types of Postganglionic Neurones Supplying Skin and Skeletal Muscle

In addition to the different types of postganglionic neurones described above, there may also be postganglionic neurones supplying such target organs as sensory receptors in the skin and skeletal muscle, extra- and intrafusal muscle fibres and capillaries and lipocytes in subcutaneous adipose tissue.

3.7.1 Sensory Receptors

Bernard (1851) described that removal of the superior cervical ganglion in the cat led to increased sensitivity of the skin of the face and cornea. In this century this observation was confirmed for the cat paw following an abdominal sympathectomy (*Dusser de Barenne* 1931) and for human patients following the same operation (*Pette* 1927; *Brown* and *Adson* 1929; *Nyström* 1949; *Tracey* and *Cockett* 1957; *Litwin* 1962). Up to 6 weeks after the sympathectomy, some of these patients developed hyperesthesias and pain from the territories which had been sympathetically denervated. The mechanism of the increased sensitivity is presently unknown.

Later, the efferent sympathetic control of sensory receptors in skin and skeletal muscle became of more general interest (*von Brücke* 1932). Led by the concept that the sympathetic nervous system prepares the organism to meet threatening stimuli from the environment (*Cannon* 1939), it seemed logical to assume that this system can increase the sensitivity of cutaneous and other receptors. Efferent control of cutaneous receptors does exist in lower vertebrates, such as frogs; stimulation of the sympathetic supply to the skin leads to activation of cutaneous receptors or to a decrease in their threshold (*Loewenstein* 1956; *Chernetski* 1964; *Spray* 1974; *Calof* et al. 1981). In humans, this point is supported by clinical observations on patients who developed pain subsequent to nerve lesions (preferentially, the median, sciatic and tibial nerves and their branches) and in whom the pain (originating from the territory of the lesioned nerve, as well as from adjacent territories in the limb concerned) could be abolished by blockade of the sympathetic activity to the painful extremity. These painful states are typical for causalgia and reflex sympathetic dystrophy (see *Livingston* 1943; *Bonica* 1953, 1979; *Richards* 1967). The conclusions from these clinical observations may be that postganglionic fibres are coupled to afferent fibres and receptors in the periphery and that postganglionic stimulation can lead to activity in primary afferent neurones, which in turn leads to pain (for literature see *Bonica* 1979; *Sunderland* 1978; *Devor* 1983; *Blumberg* and *Jänig* 1983b). This appears to be the opposite to the increased cutaneous sensitivity following sympathectomy, as described above. The questions arising are:

1. Are the receptors in skin and deep somatic tissues of mammals innervated by postganglionic adrenergic neurones with different functional characteristics than those innervating vessels?
2. Do these receptors change their response characteristics when the sympathetic neurones are activated?

Histological investigations have shown that 8%–65% of muscle spindles, but no tendon organs, are supplied by postganglionic axons. These axons were found equatorially within the capsule wall and close to the intrafusal muscle fibres at the polar regions. It is undecided whether the same postganglionic fibres supply both parts of the muscle spindle and whether in addition these fibres supply blood vessels (*Barker* and *Saito* 1981; for earlier literature see *Akoev* 1981). Electrical stimulation of sympathetic fibres leads to a small increase in activity of Ia fibres from muscle spindles followed by a silent period. It has no effect on Ib fibres from tendon organs. Generally the effects can only be observed in some of the afferents and can possibly only be elicited by electrical stimulation of the sympathetic chain at frequencies which rarely occur in vivo (*Eldred* et al. 1960; *Hunt* 1960; *Hunt* et al. 1982). Thus, it is unclear whether this innervation has biological importance.

Many cutaneous mechanoreceptors with myelinated afferents seem to be supplied also by unmyelinated fibres. Although it is believed that this additional supply is of postganglionic nature, in most cases direct proof is missing (for literature see *Akoev* 1981). The effects of sympathetic stimulation have been investigated for mechanoreceptors in cat vibrissae (*Nilsson* 1972), for guard hair receptors and type II receptors (*Pierce* and *Roberts* 1981) and for Pacinian corpuscles in the cat paw (*Freeman* and *Rowe* 1981) and in the human hand (*Hallin* and *Wiesenfeld-Hallin* 1983). Dynamic responses of slowly adapting mechanoreceptors in the vibrissae decrease, and mechanical thresholds of guard hair receptors increase. Resting activity, but not mechanosensitivity, of type II slowly adapting receptors, oriented perpendicular to the long axis of the leg (but not of those oriented longitudinally!), increases when the sympathetic supply is stimulated. The inner core of many Pacinian corpuscles is supplied by a postganglionic adrenergic axon (*Santini* et al. 1971). The neural responses of Pacinian corpuscles upon mechanical stimulation are either enhanced (*Loewenstein* and *Altamirano-Orrego* 1956; *Leitner* and *Perl* 1964; *Schiff* 1974; *Akoev* 1981; *Freeman* and *Rowe* 1981; *Hallin* and *Wiesenfeld-Hallin* 1983) or depressed (*Santini* 1976) by stimulation of the sympathetic supply or by sympathomimetic drugs. However, it cannot be excluded that the effects may be secondary to vascular changes (*Leitner* and *Perl* 1964; *Freeman* and *Rowe* 1981; *Hallin* and *Wiesenfeld-Hallin* 1983). The sympathetic chain was stimulated electrically at frequencies of 5 Hz or higher for 10 s to 3 min, i.e. conditions which probably occur very rarely, if at all, in vivo.

At present, we have almost no understanding of the biological meaning of the postganglionic sympathetic innervation of sensory structures in the skeletal muscle and skin of mammals. Several questions have to be answered and rigorously tackled experimentally: Is this innervation of sensory structures accidental and vestigial or regularly existent? Do the same postganglionic neurones supply blood vessels or other autonomic target organs as well as sensory receptors? Are receptors with unmyelinated afferents (in particular nociceptors!) also supplied by postganglionic axons?[1] Has the postganglionic supply of sensory structures perhaps a function other than the supposed change of sensitivity? What are the functional conditions which lead to an influence of postganglionic activity on sensory receptors? Other questions concern the influence of postganglionic neurones on afferent neurones from the skin and deep somatic tissues of the extremities under pathophysiological conditions, e.g. after traumata with nerve lesions. Under these conditions, afferent nociceptive neurones may be excited from the postganglionic side, e.g. by ephaptic transmission, by released noradrenaline or indirectly by neuronal dysregulation of blood flow through the peripheral tissue with subsequent tissue hypoxia (see *Blumberg* and *Jänig* 1983b).

3.7.2 Exchange Vessels and Lipocytes in Subcutaneous Adipose Tissue

Stimulation of sympathetic fibres supplying subcutaneous tissue of the canine abdominal wall induces vasoconstriction, an increase in the perme-

---

1 This problem has been attacked recently by *Roberts* and *Elardo* (1985). The authors tested primary afferent units in the saphenous nerve of cats, which were functionally identified as group III myelinated nociceptors, for their responses to electrical stimulation of the lumbar sympathetic trunk (supramaximal 10 Hz stimuli for 15 to 60 s).
The nociceptors were subdivided functionally into mechano-heat receptors, responding to both noxious heat and pressure, and high-threshold mechanoreceptors, responding only to pressure. No units were activated by sympathetic stimulation prior to sensitization by noxious heating of their receptive fields. Some 25% of the mechano-heat receptors (about 16% of the total population of nociceptors with group III fibres), which had high mechanical thresholds to stimulation with von Frey hairs, were activated by sympathetic stimulation alone after they had been sensitized by noxious heating of their receptive fields. The other nociceptive units were not activated by sympathetic stimulation. This is the *only* study which demonstrates that some cutaneous nociceptors can be activated by sympathetic stimulation. Another report shows that nociceptive afferent units with group III and IV fibres innervating the rabbit pinna were not excited and the responses of these units to pressure were not altered to electrical stimulation of the cervical sympathetic trunk (*Barasi* and *Lynn* 1983).

ability of the walls of exchange vessels (capillaries) and release of free fatty acids from lipocytes (lipolysis) (*Rosell* 1980; *Rosell* and *Belfrage* 1979). We do not know whether this also applies to the subcutaneous adipose tissue of the hindlimb and whether vasoconstriction, an increase in capillary permeability and lipolysis are induced by the activation of the same adrenergic postganglionic neurones or whether these three processes are elicited by separate populations of postganglionic neurones. At present, there is no way of recognizing postganglionic neurones which supply capillaries and lipocytes, but it must be kept in mind that they may exist as classes separate from CVC neurones.

4 Reflex Patterns of Postganglionic Neurones in Spinal Cats

A question arising immediately, when looking at the variety of the reflexes and reactions which can be elicited in functionally identified postganglionic neurones in brain-intact animals, is: To what degree are these complex reactions dependent on spinal neuronal circuits? Is it conceivable that the spinal cord contains neuronal circuits which are related to functionally different types of pre-postganglionic outputs. These circuits may form the basic building blocks of the neuronal organization of the sympathetic nevous systems in the neuraxis and may also be important for the creation of the reflex patterns observed in functionally identified postganglionic neurones. If this is true, the reflex patterns of postganglionic neurones in brain-intact animals may exist also in spinal animals, as far as the afferent inputs from skin, visceral organs and spinal canal (thermoreceptors) are concerned. Therefore, over the last 15 years, several series of experiments were conducted on chronic spinal cats (*Horeyseck* and *Jänig* 1974c; *Jänig* and *Spilok* 1978; *Jänig* and *Kümmel* 1981; *Kümmel* 1983).

4.1 Spinal Shock in Sympathetic Systems in the Cat

Transection of the spinal cord in mammals at cervical or thoracic levels is followed by severe depression of activity in sympathetic systems originating from the spinal cord below the transection. This state is called "spinal shock", at term which describes the lack of motor activity when the spinal cord is transected cranial to the motor neurones (*Hall* 1841; *Kuhn* 1950; *Guttmann* 1976a,b). The lack or decrease of ongoing and reflex activity in sympathetic neurones is reflected in an absence or weakness of responses of autonomic target organs (e.g. vascular beds of skeletal muscle and skin, sweat glands) to adequate general stimuli (e.g. change of body position, heat load). However, it is known from animal experiments (*Brooks* 1933, 1935; *Sahs* and *Fulton* 1940; *Koizumi* and *Brooks* 1972) and from clinical observations on tetra- and paraplegic humans (*Head* and *Riddoch* 1917; *Riddoch* and *Head* 1917; *Kuhn* 1950; *Guttmann* 1976a,b) that with time the spinal shock fades away and that many residual autonomic functions may have recovered in the chronic state. Investigations of functionally identified sympathetic neurones when the spinal cord, which has been isolated from the brainstem, is in a chronic state can yield information on the spinal organization of sympathetic systems. This information is not only of general biological interest but also of pathophysiological and clinical interest (*Guttmann* 1976a). The reactions and reflexes of the sympathetic systems are certainly not "normal" in chronic spinal preparations. Sprouting and other processes may occur intraspinally, and spinal neural autonomic circuits may not function with descending pathways from the brainstem and hypothalamus when the neuraxis is intact.

The recovery of ongoing and reflex activity in MVC, CVC and SM neurones in the cat, after transection of the spinal cord; i.e. "spinalization", between T_7 and L_1, varies, depending on the system tested and probably also on the type of reflex tested (*Horeyseck* and *Jänig* 1974c; *Jänig* and *Spilok* 1978). The reflexes elicited by noxious stimulation of the skin may be first seen in SM and MVC neurones 7–14 days after spinalization, in CVC neurones after 4 weeks. The reflexes elicited by innoxious stimulation of the skin (stimulation of hair follicle receptors and Pacinian corpuscles) may at first appear considerably later in these systems. The recovery of the reflexes to their final size may take 4–6 weeks for the MVC system and 7–10 weeks or longer for the CVC and the SM system. Surprisingly, the degree of resting activity in the SM system (determined by measuring frequency and size of transient resting skin potentials; *Jänig* and *Spilok* 1978) is not correlated with the time after spinalization.

Generally, the degree of ongoing and reflex activity varies considerably between individual chronic spinal cats. There may be several reasons for this. For example, the spinal neuronal machineries, once they are deprived of their descending excitatory and inhibitory neuronal control systems, may become very reactive to activation of primary afferents from the body surface and from the interior of the body. Therefore, it is not surprising to experience that the immediate experimental procedure (surgery, stimuli) may interfere with the measurements on sympathetic neurones more seriously in chronic spinal animals than in brain-intact animals. Noxious stimuli applied to the skin (by surgery) may depress ongoing activity in CVC neurones for up to several tens of minutes and may enhance activity in SM neurones for 10 min or longer (Fig. 11B,C; *Jänig* and *Spilok* 1978; *Jänig* and *Kümmel* 1981). The same may apply to the effect of visceral stimuli (see below) (*Kümmel* 1983). In tetraplegic patients, contractions of the urinary bladder, defecation, distension of the rectum and muscle spasms, which stimulate visceral afferents and afferents from skeletal muscle, may produce generalized activation of vasoconstrictor neurones and therefore a marked rise in BP (*Frankel* and *Mathias* 1976; *Guttmann* 1976a).

4.2 Vasoconstrictor Neurones Supplying Skeletal Muscle (MVC)

The ongoing activity of MVC neurones in chronic spinal animals varies considerably between animals and ranges from about 0.5 to about 3 imp/s. It is generally lower than in animals with an intact neuraxis. The percentage of postganglionic MVC neurones with ongoing activity may reach 75%–100% (*Horeyseck* and *Jänig* 1974c); however, recent investigations in our laboratory indicate that the figure most likely is lower. The ongoing activity does not show cardiac and respiratory rhythmicities (see Sects. 3.1.2, 3.1.6) and may exhibit considerable spontaneous fluctuations.

4.2.1 Stimulation of Cutaneous Receptors

MVC neurones are activated by mechanical and thermal stimulation of cutaneous nociceptors (*Horeyseck* and *Jänig* 1974c; *Jänig* 1975; *Kümmel* 1983). The activation can reach a considerable degree when compared with the reactions of MVC neurones to noxious cutaneous stimuli in brain-intact animals with intact buffer nerves. The activation declines rapidly after the stimulation period and is not followed by afterdischarges.

Stimulation of hair follicle afferents by air jets induces short-lasting depression of ongoing activity in MVC neurones in some chronic spinal cats (*Horeyseck* and *Jänig* 1974c). This reflex is identical to the corresponding one in brain-intact animals (see Fig. 3E). Other mechanical or thermal innoxious stimuli applied to the skin are ineffective in eliciting reflexes in MVC neurones.

4.2.2 Stimulation of Visceral Mechanoreceptors

Stimulation of visceral mechanoreceptors may elicit considerable activation of MVC neurones. The most powerful afferent stimuli are isovolumetric contractions of the urinary bladder and, sometimes, of the colon (Fig. 10C). Each contraction is accompanied by an activation of the MVC neurones. Surprisingly, the activation of MVC neurones following visceral stimuli is mostly smaller than that of CVC neurones (*Kümmel* and *Xu* 1983). The systemic arterial BP increases only marginally by about 2–5 mmHg, or not at all, during the MVC neurone activation. This may be explained by the fact that the vasoconstrictor neurones supplying resistance vessels of

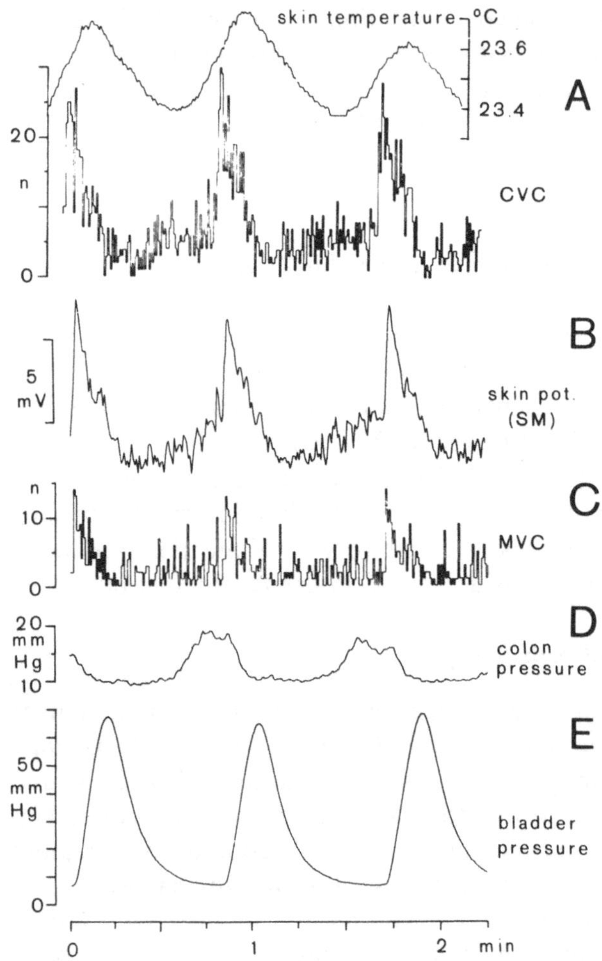

Fig. 10A–E. Reactions of CVC, SM and MVC neurones during alternating isovolumetric contractions of the urinary bladder and colon in a chronic spinal cat. The cat was spinalized at segment T_{10} 135 days before the experiment. The postganglionic activity was recorded from multifibre bundles which were isolated from the SP (hairy skin, **A**) and from a branch of the deep peroneal nerve (skeletal muscle, **C**). For the SM activity, the skin potential (*skin pot.*) was recorded from the surface of the central pad of the paw (**B**). The skin temperature was recorded on the surface of the central pad of the paw (**A**). Note that the activation of CVC neurones is followed by a decrease of the skin temperature. *n*, number of impulses per bin (0.5 s). (Modified from *Kümmel* 1983)

the upper half of the animals are still under the control of the arterial baroreceptors (*Jänig, Kümmel* and *Xu,* unpublished observation). Paraplegic humans (midthoracic interruption of the spinal cord) also show a much lower increase of BP in response to visceral stimuli (e.g. bladder stimulation) when compared to tetraplegic humans (interruption of the spinal cord at the cervical level) (*Frankel* and *Mathias* 1976; *Guttmann* 1976a,b).

4.3 Vasoconstrictor Neurones Supplying Skin (CVC)

Many postganglionic CVC neurones exhibit ongoing activity at rates of 0.1–1.5 imp/s (mean about 0.5 imp/s) in chronic spinal animals. This rate is lower than in brain-intact animals; furthermore, the percentage of CVC neurones with ongoing activity is lower in chronic spinal animals than in brain-intact animals (*Horeyseck* and *Jänig* 1974c). The ongoing activity may exhibit considerable fluctuations, but does not show cardiac and respiratory rhythmicities.

4.3.1 Stimulation of Cutaneous Receptors

Natural stimulation of innoxious cutaneous receptors has, if anything, an inhibitory influence on activity in CVC neurones in chronic spinal animals. This has been shown for stimulation of hair follicle receptors (*Horeyseck* and *Jänig* 1974c) and of Pacinian corpuscles (*Jänig* and *Spilok* 1978; *Kümmel* 1983). The latter reflex is particularly interesting; it is the same as that for CVC neurones in brain-intact cats (see Sect. 3.2.4) and the opposite of the simultaneously occurring activation of SM neurones (see Sect. 4.4.1).

Stimulation of cutaneous nociceptors (mechanically, by radiant heat) induces qualitatively the same reflexes in many CVC neurones in chronic spinal cats as in brain-intact cats: depression of ongoing activity when nociceptors of the toes of the same hindlimb are stimulated; no effect, small excitation or small depression of ongoing activity when nociceptors of the contralateral hindpaw are stimulated (Fig. 11B,C). Thus, as in the brain-intact cats, this reflex has a spatial organization (see Sect. 3.2.4) (*Horeyseck* and *Jänig* 1974c; *Jänig* and *Spilok* 1978; *Jänig* and *Kümmel* 1981). In chronic spinal animals, it can easily be shown that the increase/decrease of activity in CVC neurones to the paw are accompanied by a decrease/increase in temperature in the innervation zone on the surface of the hairless skin of the paw (Fig. 11B; *Jänig* and *Kümmel* 1981).

Certain features of this inhibitory reflex, in chronic spinal animals, are much more pronounced than in brain-intact animals (Fig. 11B,C): in many CVC units it outlasts the stimulus by minutes; the spatial organization is more pronounced. This long-lasting spinal reflex is induced by surgery in the innervation territory of the CVC neurones and may therefore prevent, or render more difficult, the recognition of postganglionic CVC neurones in chronic spinal animals (*Jänig* and *Kümmel* 1981).

4.3.2 Stimulation of Visceral Receptors and Spinal Thermoreceptors

Stimulation of receptors from the urinary bladder, colon and anus elicits excitation of most CVC neurones with ongoing activity (Fig. 10A). The

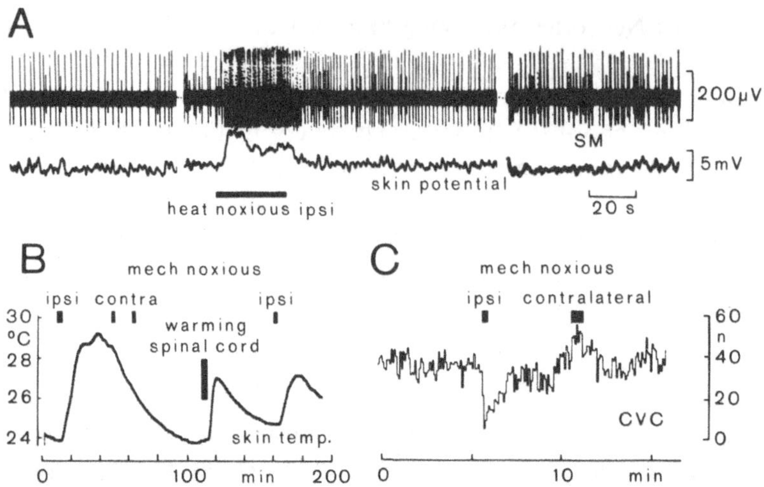

Fig. 11A–C. Reactions of SM and CVC neurones to noxious stimulation of skin in chronic spinal cats (108 days after spinalization between segments T_8 and T_{11}). **A** Recording from an SM neurone before, during and after stimulation of the toes of the ipsilateral hindpaw with radiant heat (55°C). Note the enhanced SM discharge after the stimulus. **B** Skin temperature (*temp.*) on the surface of the central pad of the left hindpaw after mechanical (*mech*) noxious stimulation of an ipsilateral toe (*ipsi*) and contralateral toe (*contra*) of the hindpaws, and after warming of the spinal cord. **C** Recording from a bundle with postganglionic CVC axons supplying the central pad of the left hindpaw. Mechanical noxious stimulation of a toe of the ipsilateral and contralateral hindpaws. (Modified from **A**, *Jänig* 1979; **B**, *Jänig* and *Kümmel* 1981; **C**, *Jänig* and *Spilok* 1978)

excitation elicited by isovolumetric contraction of the urinary bladder or colon, in the CVC neurones, is mostly stronger than in MVC neurones; otherwise, it has the same characteristics as the viscero-sympathetic reflexes in MVC neurones in chronic spinal animals (*Kümmel* and *Xu* 1983) (see Sect. 4.2.2). The reflex discharges in CVC neurones are followed by decreases in skin temperature (Fig. 10A). Commonly, the rhythmic discharges of the CVC neurones are accompanied by rhythmic fluctuations of skin temperature (*Jänig, Kümmel* and *Xu,* unpublished observations). The remarkable feature of this reflex is that it is the opposite of the inhibitory viscero-sympathetic reflex obtained in most CVC neurones in brain-intact animals (see Sect. 3.2.5).

Activity in CVC neurones, in chronic spinal animals, decreases and increases during spinal cord warming and cooling, respectively. These changes are accompanied by corresponding changes in temperature on the surface of the hairless skin of the paws (Fig. 11B). The responses to spinal cord cooling are usually smaller than those to warming (*Jänig* and *Kümmel* 1981). These responses are expected to occur in CVC neurones since it is known – at least for cats and dogs – that the spinal cord itself has principally the same thermoregulatory properties as the hypothalamus, though qualitatively not as powerful (*Thauer* 1970; *Simon* 1974).

4.4 Sudomotor (SM) Neurones

Ongoing activity in SM neurones supplying sweat glands in the hairless skin of the cat hindpaw is low or absent in chronic spinal animals (about 0.3 ± 0.2 imp/s, n = 20; *Spilok*, unpublished observations). The percentage of postganglionic SM neurones with ongoing activity is unknown and can also not be estimated from the transient fast-rising skin potentials recorded from the surface of the hairless skin of the cat paw (*Jänig* and *Spilok* 1978).

The transient ongoing skin potentials appear to occur independently on the left and right hindpaw (*Wang* 1964) whereas potentials, recorded simultaneously from lateral and medial parts of the same central pad of the cat hindpaw, are fairly well synchronized. Thus, in contrast to SM neurones supplying the left and right hindpaws, SM neurones supplying different parts of the same paw have some synchrony in their discharges. This synchrony is probably produced in the spinal cord or in the sympathetic ganglia. The innervation of lateral and medial parts of the central pad are separate, and electrical stimulation of postganglionic axons in the lateral plantar nerve does not activate sweat glands in the medial part of the central pad (which is supplied by the medial plantar nerve) and vice versa (*Jänig* and *Spilok* 1978).

Ongoing activity in SM neurones varies considerably between animals, and also in the same animal when measurements of skin potentials have been repeated on different days (*Spilok*, unpublished observations).

4.4.1 Stimulation of Cutaneous Receptors

Stimulation of Pacinian corpuscles in the hindpaw elicits excitation of SM neurones. This excitatory reflex has about the same size and characteristics as in brain-intact animals (*Jänig* and *Spilok* 1978): it cannot be elicited by stimulation of hair follicle receptors on the hindlimb and trunk by tangential air-jet stimuli nor by constant force stimuli applied to the plantar surface of the paw, which excites slowly adapting receptors. It can, however, be elicited by air-jet stimuli applied to the hairs between the toes on the plantar side. The latter stimulus excites the sensitive Pacinian corpuscles which are located in the hindpaws (*Jänig* et al. 1968). Thus, for SM neurones, this reflex is as specific in chronic spinal animals as in brain-intact animals.

Another excitatory reflex can be induced by stimulation of cutaneous nociceptors (mechanically or by radiant heat). It has about the same size as that produced by stimulation of Pacinian corpuscles. However, when compared with the corresponding reflex in SM neurones in intact animals, it is followed by an enhancement of ongoing activity that may last for up to 10 min (or sometimes longer) (Fig. 11A). This postexcitatory enhancement is not observed when Pacinian corpuscles are stimulated. It is normally accompanied by a long-lasting inhibition of activity in CVC neurones being elicited by the same noxious stimulus (Fig. 11C). The enhancement of SM activity can also be produced by electrical stimulation of the central stump of a cut cutaneous nerve with short trains. Furthermore, the first (thermal or mechanical) noxious stimulus in an individual experiment, in which no surgery has been performed on the paws, readily produces the long-lasting SM activity following the stimulus. These two latter observations argue in favour of the hypothesis that this enhancement is not due to sensitizations of cutaneous nociceptors by the stimulus (*Bessou* and *Perl* 1969; *Beck* et al. 1974; *Lynn* 1977; *Perl* et al. 1976; *Fitzgerald* 1979), but due to long-term effects in the spinal cord, which are produced by activity in primary afferent nociceptive fibres from the skin. It is also unlikely that the enhancement of ongoing SM activity occurs in the sympathetic ganglia: repetitive electrical stimulation of preganglionic axons in the LST does not lead to enhanced activity in SM neurones (*Jänig* et al. 1982; *Blumberg* and *Jänig* 1983a).

Another feature of the enhancement of ongoing activity in SM neurones is worth mentioning: during the first 4–5 weeks following transection of the spinal cord, at the thoracic level, it can barely be detected. Thereafter, when the spinal shock has faded away, it can readily be elicited in most animals. Thus, the enhancement appears when the time period of the spinal shock has passed (*Jänig* and *Spilok* 1978).

4.4.2 Stimulation of Visceral Receptors and Spinal Thermoreceptors

SM neurones in chronic spinal animals become excited when spinal visceral afferents are adequately stimulated. This applies to isovolumetric contractions of the urinary bladder and colon following distension of these organs, mechanical stimulation of receptors from the anal mucosa and mechanical stimulation of receptors in or around the vagina. The most powerful reflexes are elicited by isovolumetric contractions of the urinary bladder and the colon (Fig. 10B; each contraction is accompanied by discharges in SM neurones which occur simultaneously to the reflex discharges in CVC and MVC neurones) (*Kümmel* 1983; *Jänig, Kümmel* and *Xu* 1983).

Thermoreceptors in the spinal canal, when stimulated by warmth and cold, have no consistent effect on the SM neurones; both stimuli may lead to weak excitation of the SM system, which occurs particularly at the beginning of the stimuli (*Jänig* and *Kümmel* 1981).

5 Synthesis of Reflex Patterns Obtained in Brain-Intact and Chronic Spinal Animals

Before continuing with this review, it seems reasonable to summarize the results which have been described so far and to discuss briefly the crucial question of the functional labelling of the postganglionic neurones as MVC, CVC, etc. This summary is particularly relevant with respect to the rest of this review which describes, to a large extent, results on post- and preganglionic neurones, these having been obtained on the basis of this functional classification and which support a posteriori the functional labelling and its consequences.

5.1 Reaction Patterns of Postganglionic Neurones

When the reactions of vasoconstrictor and SM neurones to the stimuli applied to the body surface, and to the various organs of the interior of the body, are taken together, certain reaction patterns appear (Fig. 12A). These patterns have some important properties, which shall be emphasized here: from the point of view of functional classification of the postganglionic sympathetic neurones, and the tentative conclusions we are trying to draw from the reactions of these neurones, with respect to the organization of the systems in the neuraxis, it is not important whether we understand the biological meaning of all reflexes or not:

1. With respect to many afferent input systems, the reaction patterns are *reciprocally* organized between MVC and CVC neurones, on the one hand, and between CVC and SM neurones, on the other. The reciprocity may also exist between MVC and MVD neurones, with respect to their hypothalamic control, and between CVC and possibly CVD neurones, with respect to thermoreceptors in the spinal cord.

2. Activity in MVC neurones is dominated by that in the *arterial baroreceptors*. Otherwise, most afferent inputs from viscera, skin and arterial chemoreceptors have an excitatory effect.

3 Most CVC neurones are regulated in an *inhibitory mode*. The inhibition produced by stimulation of cutaneous nociceptors has a spatial organization. Some CVC neurones (supplying hairy skin) display a reaction pattern which is similar to that of MVC neurones.

4. SM neurones are activated by all afferent inputs tested so far. A unique and quite specific reflex is elicited in SM neurones by stimulation of Pacinian corpuscles (*vibration receptors*).

5. All three systems are coupled to the *respiratory system* (as revealed by the "respiratory rhythmicity" of their activity). This coupling is strong-

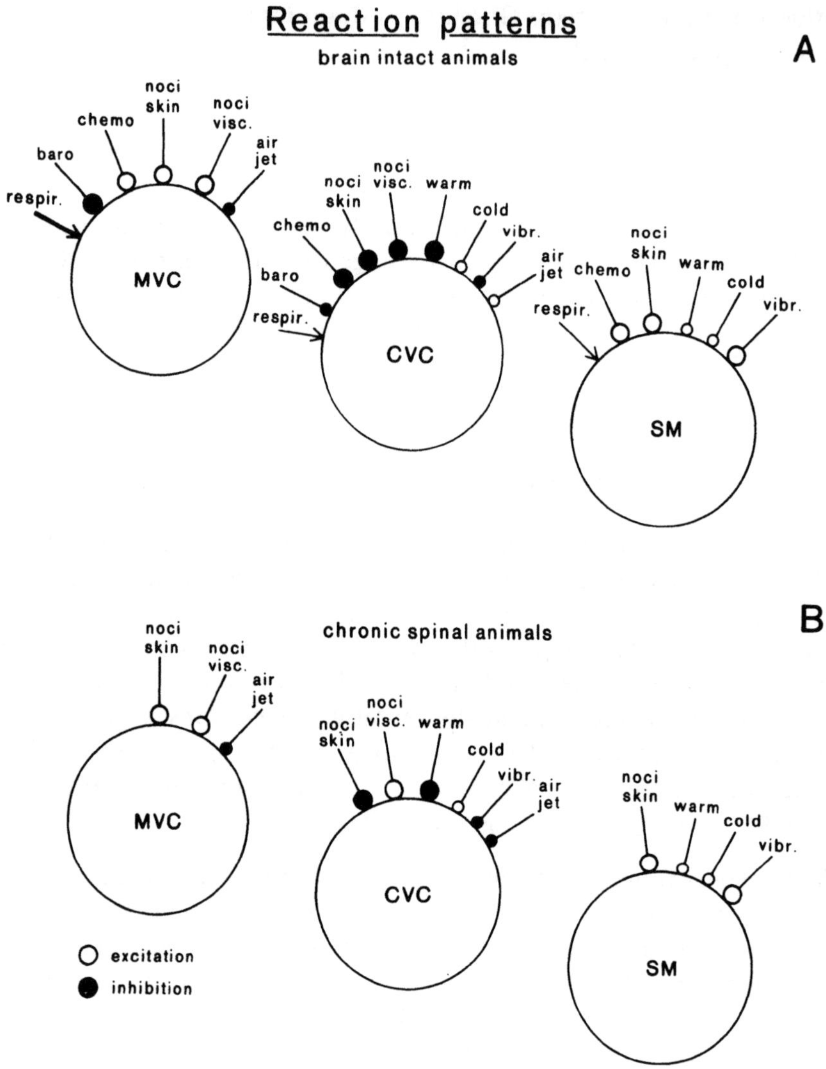

Fig. 12A,B. Reaction (reflex) patterns of vasoconstrictor neurones supplying skeletal muscle (*MVC*) and skin (*CVC*) and of SM neurones (*SM*) in chloralose-anaesthetized cats with intact neuraxis **A** and in chronic spinal cats **B**. The *open* and *closed small circles* indicate excitatory and inhibitory actions, respectively, induced by activation of the afferent input systems. *Respir.*, respiratory modulation of postganglionic activity; *chemo*, stimulation of arterial chemoreceptors; *noci skin*, stimulation of cutaneous nociceptors of the ipsilateral hindpaw; *air jet*, stimulation of hair follicle receptors; *vibr.*, stimulation of Pacinian corpuscles (vibration receptors) in the hindpaws; *warm, cold*, stimulation of thermosensitive structures in the spinal canal and/or in the hypothalamus, *visc.*, stimulation of visceral (noci) receptors in the urinary bladder and colon. (Modified from *Jänig* 1979, 1982)

est in the MVC system and weakest in the CVC and SM systems. It can be enhanced by certain experimental procedures in the latter two systems.

6. *In chronic spinal animals,* the characteristic reaction patterns are partially preserved with respect to afferent inputs from the skin and viscera (Fig. 12B). Major changes occur in the activity of CVC neurones: the inhibitory viscerosympathetic reflex, in animals with intact neuraxis (Fig. 12A), changes to an excitatory one in spinal animals (Fig. 12B). The spatial organization of the inhibitory reflex produced by stimulation of cutaneous nociceptors is enhanced, the inhibition is prolonged. SM neurones exhibit long-lasting afterdischarges following cutaneous noxious stimulation.

7. Functionally identified neurones, without ongoing activity (MVD, CVD and PM neurones), seem to be activated by very *specific* stimuli only and by most other stimuli not at all.

The existence of these reaction patterns and the distinctness of certain stimuli for activation of the silent systems argue in favour of the hypothesis that each sympathetic system has a specific central organization and that part of this is probably encoded in the neuronal circuits of the spinal cord (see Sect. 9). The reaction patterns are, furthermore, valuable and nondispensible tools for recognizing the systems with respect to function.

5.2 Functional Labelling of Postganglionic Neurones: What Justifies This?

At present, it cannot be proved in an in vivo experiment that a particular postganglionic neurone, which displays a typical reaction pattern in its activity, innervates a specific target organ. Technically, it seems impossible to record from a single postganglionic neurone with an intact connection to its target organs, to identify the neurone functionally and to stimulate it selectively in order to observe the response of the target organ. We do not know whether any target organ of the sympathetic postganglionic neurones will respond to the stimulation of a *single* postganglionic neurone. Thus, for the time being, and probably also in the future, we have no reliable method available for the identification of postganglionic (not to mention preganglionic) neurones with respect to their target organs. The closest one can get, at present, is by recording activity in postganglionic neurones and effector responses simultaneously. This is a relatively reliable method for the SM system since the negative transient skin potentials, on the surface of the hairless skin, are a fast indicator of the activation of the sweat glands (Fig. 8A–C; *Jänig* and *Kümmel* 1977). For most autonomic effector responses, this close temporal relation between neuronal responses and responses of the target organs does not exist. Thus, the labelling of postganglionic neurones such as MVC and CVC is based on a *correlation* between their discharges and the responses of target organs and on the reaction patterns observed in these neurones. This functional labelling of the neurones provides an at least reasonable classification of them, which is fully supported by the observation that different types of nerves to the skin and skeletal muscle of the cat hindlimb and tail contain, as seen from the target organs, different compositions of postganglionic axons (see Table 1). The functional labelling, however, does not necessarily mean that a particular postganglionic neurone actually innervates the target organ it is supposed to, but one may assume that, statistically, the probability of the neurone doing so is very high.

6 Reflex Patterns in Preganglionic Neurones

Postganglionic neurones which supply different target organs exhibit different reaction patterns (Fig. 12). Do preganglionic neurones, which make synapses with the postganglionic neurones supplying the skin and skeletal muscle of the cat's hindlimb, exhibit the same reaction patterns? In other words, are these reaction patterns faithfully transmitted from pre- to postganglionic neurones as if there were separate channels (Fig. 1)? The basic question concerning the specificity of synaptic connections between pre- and postganglionic neurones was first discussed by *Langley* (1892, 1903). He stimulated the preganglionic sympathetic outflow from different thoracic segments to the superior cervical ganglion in the cat, dog and rabbit and observed the responses of the sympathetic target organs. He found that stimulation of each of the first seven thoracic spinal nerves produces a different pattern of autonomic responses. For example, stimulation of the first nerve leads to pupillary dilation and retraction of the nictitating membrane, but little or no vasoconstriction of the ear, whereas stimulation of the T_4 spinal nerve causes vasoconstriction of the ear and piloerection, but has little effect on the eye. Langley's experiments, which much later were repeatedly confirmed (*Murray* and *Thompson* 1957; *Guth* and *Bernstein* 1961; *Njå* and *Purves* 1977; see also *Purves* and *Lichtman* 1978), indicate that there is specificity in the innervation of postganglionic sympathetic neurones by preganglionic axons, with respect to the target organs. Thus, as seen from Langley's experiments, one should expect that preganglionic neurones display the same reaction patterns as postganglionic neurones and also that these neurones can therefore be labelled functionally. How the specificity in the connection between pre- and postganglionic neurones has been brought about we do not yet know (for discussion of this point see *Purves* and *Lichtman* 1978).

6.1 Ongoing Activity in Preganglionic Neurones

The majority of preganglionic axons in the LST of the cat, which project distally to ganglion L_5, probably synapse with postganglionic neurones supplying the skin and skeletal muscle of the hindlimb, tail and lower trunk; some may supply viscera (*Langley* and *Anderson* 1894; *Kuo* et al. 1984; for labelling of ganglia of the lumbosacral sympathetic trunk, see Sect. 7.1 and *Baron* et al. 1985a). About 27% (115 of 426 neurones; see *Jänig* and *Szulczyk* 1980) of these preganglionic neurones from the spinal segments L_2 and L_3 have ongoing activity. This percentage is almost identical to that obtained on preganglionic neurones projecting to the superior cervical ganglion (*Jänig* and *Schmidt* 1970). The rate of ongoing activity ranges from 0.1 to about 4.8 imp/s (*Jänig* and *Szulczyk* 1980). Other authors have obtained similar rates of ongoing activity for lumbar preganglionic neurones (0.1–4 imp/s, mean 2.1 imp/s by *Sato* 1972; 0.1–11 imp/s, mean 1.8 imp/s by *Seller* 1973; 0.1–9 imp/s, mean 1.4 imp/s by *Kaufman* and *Koizumi* 1971). Preganglionic neurones with the reaction patterns of MVC neurones have a higher rate of ongoing activity (mean ± 1 SD, 1.8 ± 1.3 imp/s, n = 26) than preganglionic neurones with the reaction patterns of CVC neurones (0.9 ± 0.6 imp/s, n = 47). From the low rate of ongoing activity in postganglionic SM neurones and the absence of it in the other types of postganglionic neurones (PM, MVD, CVD; see Table 2a), one would expect that the corresponding preganglionic neurones also have little or no ongoing activity.

The rates of resting activity in preganglionic lumbar neurones are similar to those in thoracic preganglionic neurones, which project into the cervical sympathetic trunk. The mean values range in various investigations from 1.4 to 2.9 imp/s (*Polosa* 1968; *Jänig* and *Schmidt* 1970; *Polosa* and *Wyszogrodski* 1973; *Mannard* and *Polosa* 1973). Only *McLachlan* and *Hirst* (1980) observed lower values (about 0.9 imp/s). The highest values observed were around 5–7 imp/s. Again, the median values were close to 1 imp/s (*Jänig* and *Schmidt*, unpublished observations).

6.2 Functional Identification of Preganglionic Neurones

The majority of the preganglionic neurones with resting activity from segments L_2 and L_3, which project with their axons into the LST distal to ganglion L_4, can be classified in groups having similar discharge patterns as postganglionic MVC, CVC and SM neurones. Neurones which are excited by stimulation of arterial chemoreceptors exhibit a strong cardiac rhythmicity in their activity, and some of them are excited by stimulation of cutaneous nociceptors. This pattern is exactly the same as that in postganglionic MVC neurones (Fig. 12A). Other preganglionic neurones exhibit no or only a weak cardiac rhythmicity in their activity; these neurones are inhibited by stimulation of arterial chemoreceptors and/or cutaneous nociceptors. The latter preganglionic group has a reaction pattern one would expect for CVC neurones (Fig. 12A). A few preganglionic neurones could be activated by stimulation of Pacinian corpuscles but displayed no cardiac rhythmicity in their activity. These neurones were probably preganglionic SM neurones (*Jänig* and *Szulczyk* 1980). Of 115 preganglionic neurones that were tested for their reaction pattern, 23% behaved in the same manner as MVC neurones, 42% as CVC neurones and 5% as SM neurones. Thirty percent of the neurones could not be classified; they did not, however, exhibit any pattern contradictory to the patterns described above. Thus, it seems that at least for the MVC, CVC and SM systems, the information may be transmitted by separate pre-postganglionic channels to the respective target organs. Therefore, the reaction patterns obtained on the postganglionic side (Fig. 12A) can also be used as functional labels enabling recognition of preganglionic neurones, with respect to the target organs.

6.3 Silent Preganglionic Neurones

Seventy-three percent of the neurones which project, through WRL_2 and WRL_3, into the LST to ganglion L_5 and more distal ganglia have no ongoing activity. Most of these silent neurones cannot be activated by any of the natural stimuli used (*Jänig* and *Szulczyk* 1980). Similar percentages of neurones without ongoing activity and reflex activity project into the cervical sympathetic trunk (*Polosa* 1968; *Jänig* and *Schmidt* 1970).

Only about 8% of these silent neurones which project into the LST are afferent. This percentage is evaluated from histological studies in which HRP was applied to the central stump of the LST between ganglia L_5 and L_6 and in which labelled afferent and preganglionic cell bodies were counted (*Jänig* and *McLachlan*, unpublished observations). After deducting the 8% afferent neurones from the total population of silent neurones, still about 70% of all neurones which project into the LST to ganglion L_5 and more distally are silent, and these must be preganglionic.

Part of the silent preganglionic neurones innervate PM, MVD and CVD neurones. These latter neurones are also silent and cannot be activated by natural peripheral stimuli. The numbers of these types of postganglionic neurones are probably low compared to the total population of postganglionic neurones supplying skin and skeletal muscle; therefore, it appears likely that only some silent preganglionic neurones innervate PM, MVD and CVD neurones.

Is the observation of a high percentage of preganglionic neurones without ongoing and reflex activity an experimental artefact or is it real? There are certainly some technical reasons: First, the chloralose anaesthesia used may depress ongoing and reflex activity. Though this seems obvious, it is not self-evident; it has to be proven for each sympathetic system separately (see Sect. 8.1). Second, isolation of strands with single or few preganglionic axons, from the LST, may lead to a selection of the thicker myelinated axons since these produce bigger spikes. Since fewer preganglionic neurones with thick myelinated axons (which conduct at more than 5–8 m/s) have resting and reflex activity than neurones with thin myelinated and unmyelinated axons (*Jänig* and *Schmidt* 1970; *Jänig* and *Szulczyk* 1980), one would expect the sampling technique to result in a higher proportion of silent preganglionic axons. This point is supported by histological investigations of the WR T_7–T_9 in the cat, which show that about 40% of preganglionic axons in these rami are unmyelinated (*Coggeshall* et al. 1976; *Coggeshall* and *Galbraith* 1978), yet the percentage of sampled unmyelinated preganglionic axons (conduction velocity below 2 m/s) is always considerably lower than 40% (*Jänig* and *Szulczyk* 1980).

Keeping the technical limitations in mind, it may also well be that many silent preganglionic neurones which conduct with their axons at less than 10 m/s (see Fig. 15) synapse with postganglionic vasoconstrictor neurones supplying skeletal muscle and skin and with postganglionic SM neurones. These preganglionic neurones may be activated during specific functional states of the organism, e.g. during emergency, severe exposure to cold and hypoxia and during special emotional states of the organism. The activation of the silent preganglionic neurones may particularly be dependent on the functioning of higher brain structures, such as the hypothalamus and limbic system. This interpretation implies that the postganglionic neurones function as final common paths in the sense that they receive functionally different convergent preganglionic inputs.

7 Location, Quantitative Characteristics of Pre- and Postganglionic Neurones and Conduction Velocities of Pre- and Postganglionic Axons

The following section concentrates on some quantitative static properties of pre- and postganglionic neurones, which were obtained with neurophysiological and morphological methods. These static properties, such as

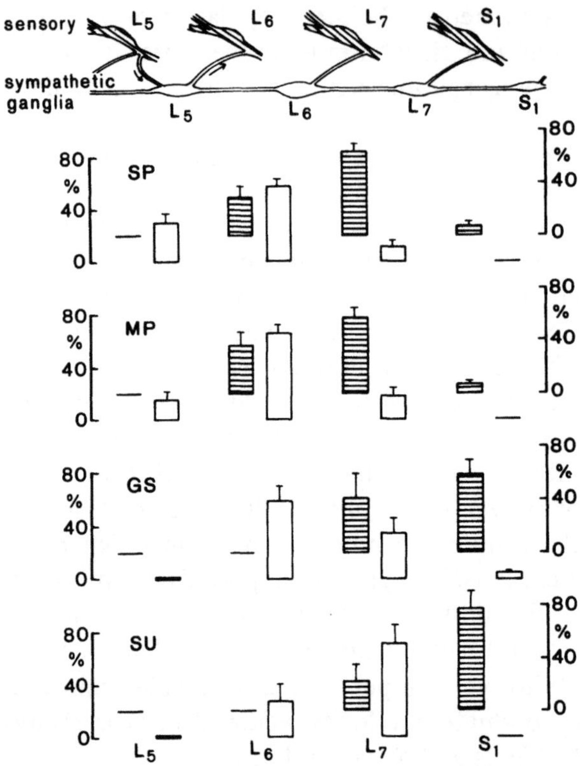

Fig. 13. Mean percentage distributions, at different segmental levels, of postganglionic sympathetic (*white columns, left ordinate scale*) and afferent neurones (*shaded columns, right ordinate scale*) projecting into hindlimb nerves of the cat. HRP solution was applied to freshly cut transected axons of the nerves, and somata labelled retrogradely in sympathetic paravertebral ganglia and spinal ganglia were counted. *SP*, $n = 11$; *MP*, $n = 5$; *GS*, $n = 5$; *SU*, $n = 5$. The *bars* indicate 1 SEM. *WR*, white ramus; *GR*, grey ramus. The percentages of labelled sympathetic cells found in paravertebral ganglia L_3 and L_4 or L_4 were 1.2 ± 0.3% (SP), 0.4 ± 0.2% (MP), 0.2 ± 0.1% (GS) and 0.2 ± 0.2% (SU) (mean ± 1 SEM). (Modified from *McLachlan* and *Jänig* 1983 and *Jänig* and *McLachlan* 1984)

conduction velocities of the axons, size and location of somata, and number of postganglionic neurones projecting into skin and muscle nerves, supply valuable information about the sympathetic systems, which in turn supplements the results on the discharge properties of sympathetic neurones.

7.1 Segmental Location of Postganglionic Somata with Respect to Afferent Somata

In the lumbosacral region, the first four to five lumbar spinal nerves send WR to the corresponding ganglia of the LST. These are then labelled according to the origin of the WR ganglia L_1 to L_5 of the sympathetic trunk. Each ganglion of the LST sends its grey rami to the next lower spinal nerve. Thus, ganglion L_5 sends its grey rami to the spinal nerve L_6 etc. (see upper part in Fig. 13; this kind of labelling is somewhat at variance with the literature and has been discussed in *Baron* et al. 1985a).

The majority (about 90% or more) of the somata of postganglionic neurones, which project into a hindlimb nerve, are situated in two adjacent ganglia of the ipsilateral LST. These ganglia send their grey rami to those two spinal nerves which contain 90%–100% of the afferent axons of the same hindlimb nerve. Each nerve has a characteristic distribution of postganglionic and afferent somata (Fig. 13). The remaining postganglionic somata lie in adjacent sympathetic ganglia; almost no postganglionic somata (less than 0.6%) lie in contralateral sympathetic ganglia. Thus, the sympathetic postganglionic outflow to the skin and skeletal muscle of the cat hindlimb is almost as precise in its spatial organization as the afferent system (*McLachlan* and *Jänig* 1983).

It seems very unlikely (although it cannot be completely ruled out) that, at least for the cat, afferents from the hindlimb pass through the lumbosacral sympathetic trunk and have their cell bodies in dorsal root ganglia, these being situated cranial to the dorsal root ganglia L_5 to L_7 or in the sympathetic trunk itself (see *Pick* 1970). In the first case, considerable numbers of stained afferent cell bodies should be found in spinal ganglia which are located cranial to the spinal ganglia with labelled cells when HRP is applied to a hindlimb nerve (see Fig. 13; *McLachlan* and *Jänig* 1983). In the second case, labelled cell bodies without dendrites should be seen in ganglia of the sympathetic trunk. Both cases have not been observed. The first case may be at variance with the earlier work of *Kuntz* (*Kuntz* and *Farnsworth* 1931; *Kuntz* and *Saccomanno* 1942; *Kuntz* 1951). He found that myelinated afferent axons in cats run from the spinal ganglia L_1 to L_5 through the sympathetic chain and through the grey rami L_6 to S_1 to the lumbosacral spinal nerves. He believed that these afferent fibres or some of them innervate blood vessels of the hindlimb.

7.2 Segmental Distribution of Pre- and Postganglionic Neurones with Respect to Different Types of Target Organs

Since *Langley* (1903), very little work has been done on the segmental distribution of preganglionic neurones which innervate postganglionic neurones supplying the hindlimb and tail (Fig. 14A). Preganglionic axons innervating postganglionic CVC neurones to the hairless skin of the paw originate mainly from spinal segments L_1 to L_3 (*Langley* 1891b, 1894/95). Preganglionic axons innervating postganglionic SM neurones have the same

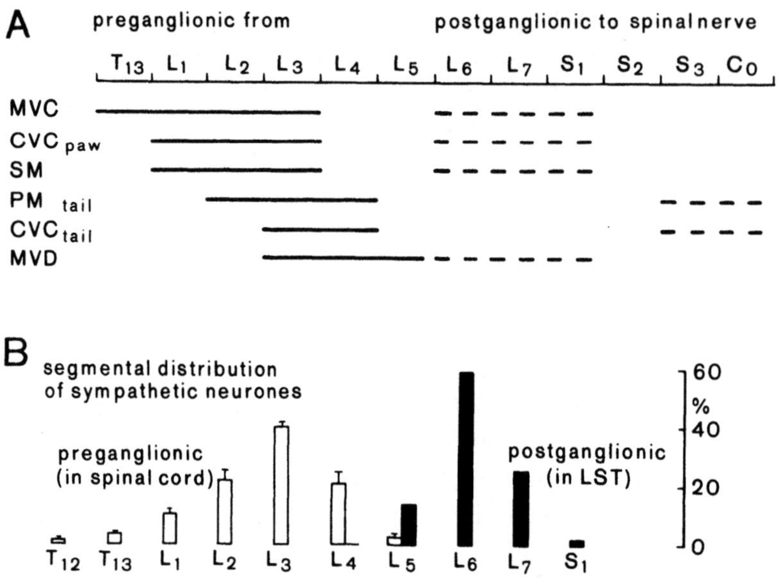

Fig. 14. A Main locations of preganglionic neurones in spinal segments and main projection of postganglionic neurones in spinal nerves with respect to different target organs of hindlimb and tail. Note that most postganglionic neurones projecting into spinal nerve L_6 are situated in the paravertebral ganglion L_5, etc. (see Fig. 13). For details see text. **B** Segmental distribution of preganglionic neurones which project into the sympathetic trunk distal to the most caudal lumbar splanchnic nerve (*open columns*) and of postganglionic neurones in the sympathetic trunk ganglia which project into the distal part of the hindlimb (*black columns*) in the cat. The segmental distribution of the preganglionic neurones was determined from preparations in which HRP solution was applied to the freshly cut LST between ganglia L_5 and L_6 (mean + 1 SEM of five preparations; *Jänig* and *McLachlan*, unpublished observations). The segmental distribution of the postganglionic neurones was evaluated from *McLachlan* and *Jänig* (1983). It was assumed that the innervation of the distal hindlimb by postganglionic fibres is equivalent to three times the innervation of the gastrocnemius soleus muscles, two times that in the SP, two times that in the MP and one times that in the SU. The nerves were weighted according to their content of postganglionic axons (see also Fig. 13)

origin (*Langley* 1894/95). Preganglionic axons innervating postganglionic PM and CVC neurones to the tail originate from spinal segments L_2 to L_4 and from L_3 to L_4, respectively (*Langley* and *Sherrington* 1891; *Langley* 1894, 1894/95). Preganglionic axons innervating postganglionic MVC neurones of the hindlimb originate from spinal segments T_{13} to L_3 (*Sonnenschein* and *Weissman* 1978). Preganglionic axons innervating postganglionic MVD neurones originate from spinal segments L_3 to L_5 (*Sonnenschein* and *Weissman* 1978). Thus, as with the spatial organization of the preganglionic sympathetic outflow to the head (*Langley* 1903), it is likely that there is a similar, though more overlapping, organization of the preganglionic sympathetic outflow to different target organs of the hindlimb and tail.

The respective postganglionic somata are situated in ganglia L_5 to L_7 for the hindlimb and in ganglia S_2 and S_3 for the tail. They project into the spinal nerves L_6 to S_1 and S_3 to Co, respectively (Fig. 14A).

The physiologically evaluated segmental distributions of pre- and postganglionic sympathetic neurones regulating target organs of the hindlimb correspond to the segmental distributions of pre- and postganglionic somata in the spinal cord and in the sympathetic trunk, respectively, which were determined morphologically. Most postganglionic somata lie in ganglia L_5 to L_7 (Fig. 14B, black columns; see also Fig. 13). Most preganglionic somata lie in the spinal segments T_{13} to L_4 (Fig. 14B, open columns). The latter was determined from experiments in which preganglionic neurones were labelled by HRP. HRP was applied to freshly cut preganglionic axons in the LST distal to the most distal lumbar splanchnic nerve (between ganglia L_5 and L_6). In this way, labelling of visceral preganglionic neurones, which project into the inferior mesenteric ganglion, was avoided (*Baron* et al. 1985a). The preganglionic axons in the LST distal to the most distal lumbar splanchnic nerve should only synapse with postganglionic neurones which project into lumbar spinal nerves L_6 and L_7 and into sacral and Co spinal nerves. A few preganglionic neurones may synapse with postganglionic neurones which project into the pelvic nerves (*Kuo* et al. 1984).

Morphological studies conducted recently have shown that preganglionic neurones which project in the LST to ganglion L_6 and more distally have a very distinct spatial distribution in the spinal cord compared to visceral preganglionic neurones which project in the lumbar splanchnic nerves. Most cell bodies of the preganglionic neurones projecting in the LST are situated in the pars funicularis (white matter) and pars principalis of the classical intermediolateral cell column; some cell bodies are situated more medially, the latter are found particularly in lumbar rostral segments. Most visceral preganglionic cell bodies lie somewhat medial to the classical intermediolateral cell column and practically no lumbar visceral preganglionic neurones are found in the white matter (*Baron* et al. 1985b; *Jänig* and *McLachlan*, unpublished observation). These morphological findings indicate that sympathetic preganglionic neurones may be organized, according to their functions, in spatially distinct groups in the lumbar spinal cord.

7.3 Conduction Velocities of Pre- and Postganglionic Axons

One important factor in support of the idea that different target organs in skeletal muscle and skin are supplied by separate pre-postganglionic channels would be if one could show that pre- and postganglionic axons belonging to these functionally separate channels conduct differently.

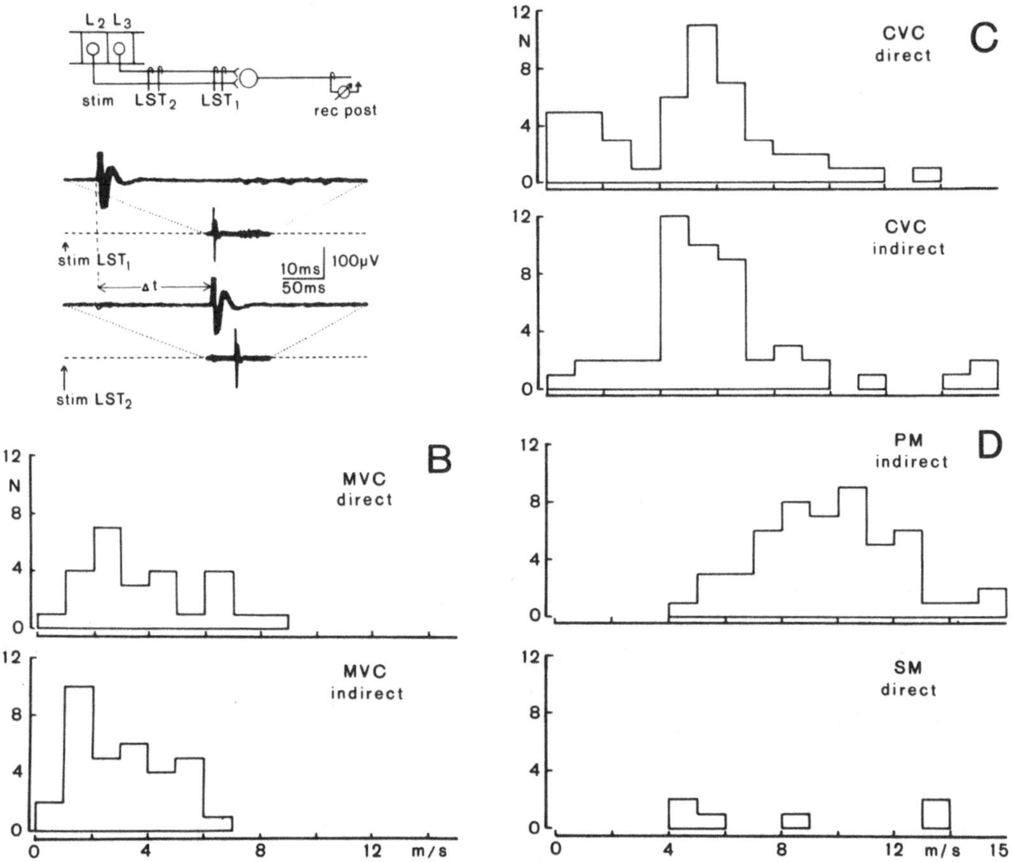

Fig. 15A–D. Distribution of conduction velocities of different types of preganglionic axons. The conduction velocities were determined either *directly* from functionally identified preganglionic neurones (see Fig. 2C) or *indirectly* as outlined in **A**. **A** The activity was recorded from the postganglionic neurones (*rec post*), which were mostly functionally identified. The preganglionic axons in the LST were stimulated at two sites being ΔS apart (*stim* LST_1, LST_2) at strengths of about twice the threshold for exciting the postganglionic neurone. The latencies of the postganglionic neurone were determined from 50–200 single responses. The distance between both pairs of preganglionic stimulation electrodes (Δs) divided by the latency difference Δt of the median responses in the postganglionic neurones yields information as to the conduction velocities of preganglionic axons. Note that distributions of conduction velocities obtained with the direct and indirect method for preganglionic MVC and CVC axons are very similar. (Modified from *Jänig* and *Szulczyk* 1980, 1981)

Previously, *Bishop* and *Heinbecker* (1932) indicated that electrical stimulation of pre- and postganglionic axons supplying the cat head leads to contraction of the nictitating membrane and dilatation of the pupil at low stimulus strength and that thinner myelinated preganglionic fibres and unmyelinated postganglionic fibres have to be stimulated in order to produce a constriction of blood vessels in the eye and ear. Later, it was shown that the nictitating membrane is innervated by thin myelinated postganglionic axons (*Kosterlitz* et al. 1964). These experiments showed that the conduction velocities of pre- and postganglionic axons, to blood vessels and non-vascular target organs of the eye, may be different. The following section reports the conduction velocities of pre- and postganglionic axons supplying the skeletal muscle and skin of the hindlimb and some morphological parameters of postganglionic neurones.

7.3.1 Preganglionic Neurones

The distributions of the conduction velocities of axons of functionally identified preganglionic neurones are different for each of the three types of neurones identified. Preganglionic axons exhibiting the same reaction patterns as postganglionic MVC neurones conduct at 3.9 ± 2.2 m/s and are the slowest (Fig. 15B, direct). Preganglionic axons exhibiting the same discharge properties as prostganglionic CVC neurones conduct at 6.6 ± 2.2 m/s (myelinated axons) and at 1.3 ± 0.6 m/s (unmyelinated axons; Fig. 15C, direct). Preganglionic axons, which can be activated in the same manner as postganglionic SM neurones by vibrational stimuli, conduct at 4.5–18.5 m/s (Fig. 15D). The distribution of the values for the preganglionic MVC and CVC axons corresponds quite well to the distribution of values which were obtained with an indirect method (Fig. 15A). The recording was made from single postganglionic axons, and the sympathetic trunk was stimulated successively at two different sites (LST_1 and LST_2 in Fig. 15A). The distance between the stimulation electrodes (ΔS) and the latency differences between the responses of a postganglionic neurone (Δt) give a measure of the conduction velocity of the preganglionic axons synapsing with the postganglionic neurone recorded from. The distributions of the conduction velocities of preganglionic axons obtained with this technique are practically identical to those obtained directly from the preganglionic axons for CVC and MVC systems (see Fig. 15B,C; *Jänig* and *Szulczyk* 1980). Preganglionic axons converging on postganglionic PM neurones conduct the fastest when measured with the indirect method (Fig. 15D; 9.9 ± 2.9 m/s; *Grosse* and *Jänig* 1976). These measurements clearly show that the distribution of the size of axons, and probably also of somata of each functional type of preganglionic neurone, varies.

About 70% of the preganglionic neurones which project into the LST distal to ganglion L_4 are silent and cannot be classified (see Sect. 6.3; difference between solid line diagram and dashed line diagram in Fig. 16A). The conduction velocity of these silent preganglionic axons ranges from 0.5 to more than 15 m/s; the proportion of silent preganglionic fibres which conduct at more than 6–8 m/s is higher than in the population of preganglionic axons with ongoing activity. Many of these fast conducting

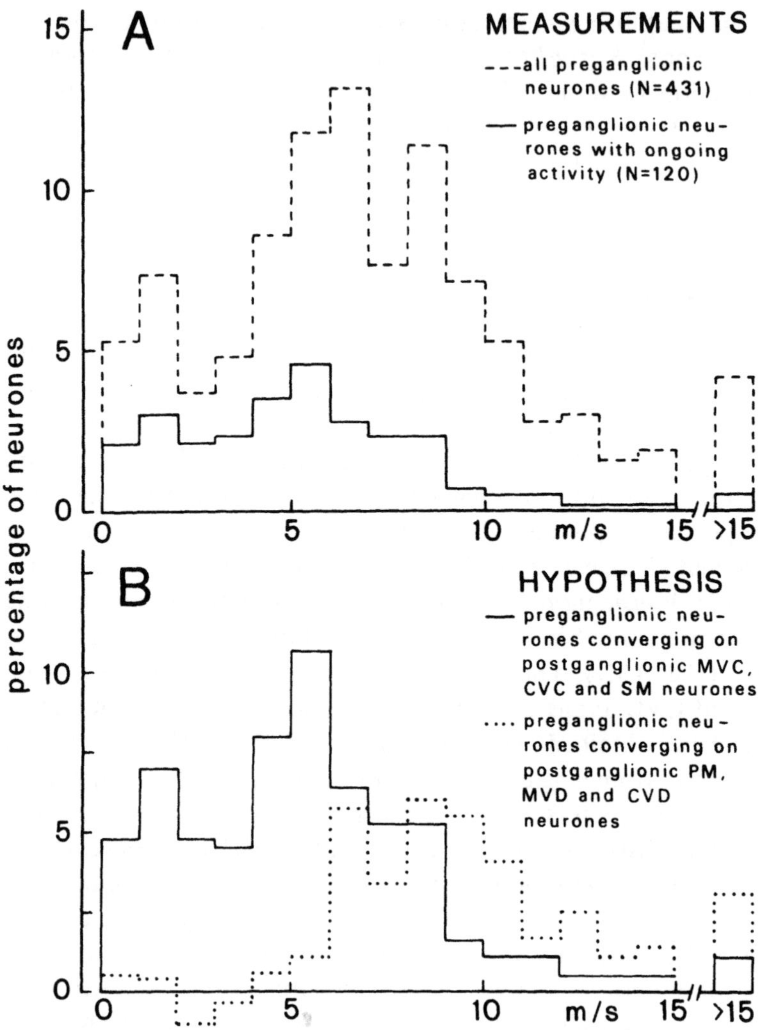

Fig. 16A,B. Percentage distributions of conduction velocities of axons of neurones projecting through WR L_2 and L_3 in the LST distal to paravertebral ganglion L_4. **A** Preganglionic neurones with ongoing activity (*solid line histogram*) and all axons which were isolated from the LST (including those with ongoing activity). Some 5%–10% of the silent axons may have been afferent. **B** The assumptions were made that postganglionic MVC, CVC and SM neurones obtain input from preganglionic neurones with ongoing activity and from silent preganglionic neurones and that the distribution of the conduction velocities of these converging silent preganglionic axons is similar to that of the preganglionic axons with ongoing activity. On the basis of these assumptions, about 65% of the preganglionic neurones would converge on postganglionic MVC, CVC and SM neurones (*solid line histogram,* normalized to *broken line histogram* in **A**) and about 35% on silent postganglionic neurones (*dotted line histogram,* difference between *broken line histogram* in **A** and *solid line histogram* in **B**). Most preganglionic axons of the latter group conduct at more than 6 m/s. (**A,** *Jänig* and *Szulczyk* 1980)

preganglionic axons most certainly do not converge on postganglionic CVC and MVC neurones (see Fig. 15), but probably on postganglionic PM, MVD and CVD neurones. However, it is very unlikely that all silent preganglionic neurones converge on postganglionic PM, MVD and CVD neurones. If one assumes that the conduction velocities of the silent preganglionic axons which synapse with postganglionic MVC, CVC and SM neurones have the same distribution as the conduction velocity of preganglionic axons with ongoing activity, then about two-thirds of the preganglionic axons would converge on the latter postganglionic neurones (solid line diagram in Fig. 16B) and only one-third on postganglionic PM, MVD and CVD neurones (dotted line diagram in Fig. 16B). The conduction velocities of the axons of these preganglionic neurones are largely above 6 m/s, and their distribution is very similar to the distribution of the conduction velocities of preganglionic PM axons (compare dotted line diagram in Fig. 16B with upper solid line diagram in Fig. 15D).

7.3.2 Postganglionic Neurones

Postganglionic axons in nerves supplying the hairy skin of the hindlimb conduct, on average, more slowly than afferent unmyelinated axons (Fig. 17A; *Grosse* and *Jänig* 1976; *Blumberg* and *Jänig* 1982). For example, postganglionic axons in the sural and superficial peroneal nerves conduct at 0.67 ± 0.13 m/s (mean ± 1 SD; n = 481) and unmyelinated afferent axons at 0.87 ± 0.16 m/s (n = 628). This difference between postganglionic neurones and cutaneous afferent neurones with unmyelinated axons can be seen in the size of the unmyelinated axons in the peripheral nerves and of the somata in the respective ganglia. The size of postganglionic axons is, on average, smaller than the size of unmyelinated afferent axons in the sural nerve (Fig. 17B; *Hoffmeister* and *Jänig*, unpublished observations), and somata of postganglionic neurones supplying skin are, on average, smaller than somata of afferent neurones with unmyelinated fibres supplying skin (Fig. 17C; *McLachlan* and *Jänig* 1983).

The above seems not to apply to muscle nerves. Somata of postganglionic neurones supplying the gastrocnemius-soleus muscle are, on average, larger than somata of afferent neurones with unmyelinated axons supplying the same muscles (Fig. 17D; *McLachlan* and *Jänig* 1983).

The distributions of conduction velocities in axons of functionally identified postganglionic neurones also vary (Fig. 18). CVC axons are the slowest conducting postganglionic axons (0.59 ± 0.13 m/s). MVC (0.79 ± 0.22 m/s), PM (0.84 ± 0.14 m/s) and SM axons (0.81 ± 0.1 m/s) conduct faster than CVC axons (*Grosse* and *Jänig* 1976; *Jänig* and *Kümmel* 1977; *Jänig* and *Szulczyk* 1981). MVD axons seem to conduct faster than MVC axons; howerver, exact measurements are not available (*Horeyseck* et al. 1976).

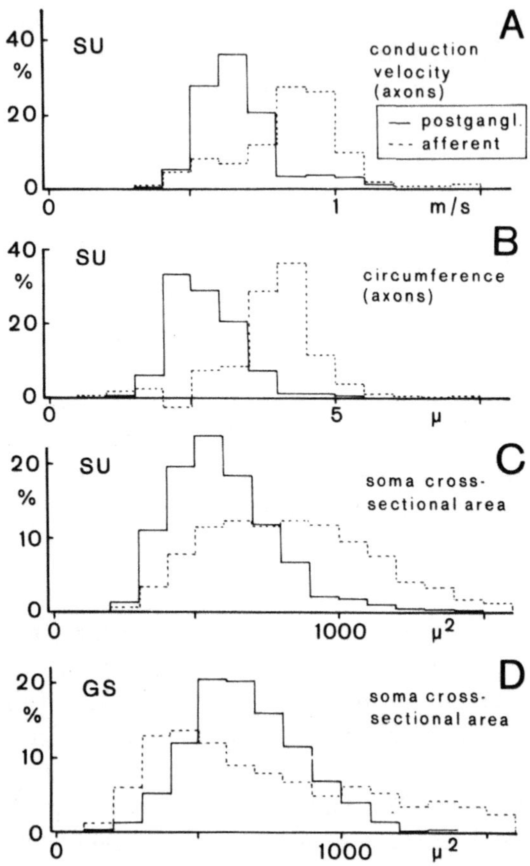

Fig. 17A–D. Percentage distributions of conduction velocities and of size of unmyelinated (postganglionic and afferent) axons in the SU and of soma size of postganglionic and small afferent neurones projecting into the SU. **A** Distribution of conduction velocities. Values from five experiments. Postganglionic axons identified by stimulation of the preganglionic axons in the LST (mean ± 1 SD, 0.65 ± 0.13 m/s, n = 275). Afferent axons identified per exclusion (mean ± 1 SD, 0.85 ± 0.22 m/s, n = 261). (*Hoffmeister* and *Jänig*, unpublished observations). **B** Distribution of circumferences of unmyelinated axons measured electron-microscopically. Values from two experiments. Size of postganglionic profiles determined 21 days after deafferentation (n = 176). Size distribution of afferent axons is the difference between size distribution of unmyelinated axons in contralateral (control) SU (n = 569) and size distribution of postganglionic axons. It was assumed that 29% of the unmyelinated axons in the SU are postganglionic (see Table 4, last line). (*Hoffmeister* and *Jänig*, unpublished observations). **C** Distributions of soma cross-sectional areas of postganglionic neurones and small afferent neurones labelled retrogradely with HRP from the SU (one experiment, $n \geqslant 400$ somata per class of neurone). (*McLachlan* and *Jänig* 1983). **D** Distribution of soma cross-sectional areas of postganglionic neurones and small afferent neurones labelled retrogradely with HRP from the nerves to the gastrocnemius-soleus (*GS*) muscles (one experiment, $n \geqslant 400$ somata per class of neurone). (*McLachlan* and *Jänig* 1983)

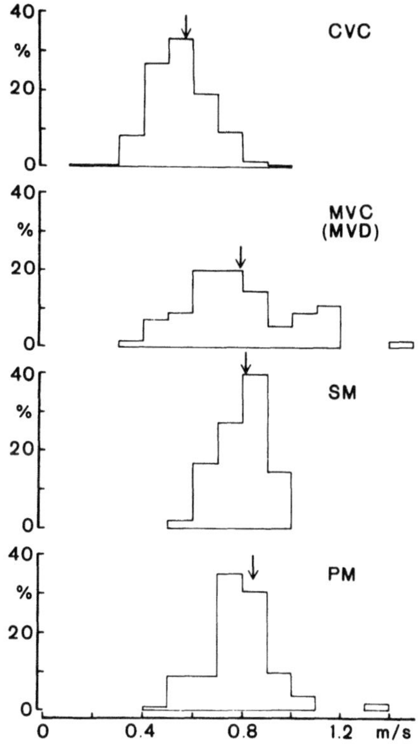

Fig. 18. Percentage distribution of conduction velocities of different types of postganglionic axons at 32°–34°C. (For temperature correction, see *Blumberg* and *Jänig* 1982). The *arrows* indicate the means. *CVC*, $n = 193$; *MVC, MVD* (not discriminated), $n = 55$; *SM*, $n = 48$; *PM*, $n = 111$. (Modified from *Jänig* and *Szulczyk* 1981 and *Jänig* and *Nelke*, unpublished observations)

The difference in conduction between different types of functionally identified postganglionic axons is usually (but not always!) reflected in the size and rise time of the action potentials recorded extracellularly from the axons. Action potentials recorded from SM, PM and MVD axons are larger in size and have a faster rise time than action potentials recorded from CVC and MVC axons (*Grosse* and *Jänig* 1976; *Horeyseck* et al. 1976; *Jänig* and *Kümmel* 1977)

7.4 Numbers of Postganglionic and Afferent Neurones Projecting into Nerves to Skeletal Muscle and Skin of the Cat Hindlimb

Estimates of the total numbers of postganglionic and afferent neurones projecting into various nerves of the cat hindlimb have recently been made from studies in which cell bodies of sensory and postganglionic axons in the nerves have been labelled retrogradely with HRP. Labelled somata in ganglia of the dorsal root and sympathetic trunk, and myelinated fibres in the peripheral nerves, were counted and the following assumptions made:

Table 4. Composition of muscle and skin nerves of the cat hindlimb

A. Numerical estimates	GS	MP	SU	SP
Skeletomotor (myelinated)	900			
Sensory, myelinated	700	2400	800	3900
Sensory, unmyelinated	1100	3200	3900	9000
Postganglionic, unmyelinated	3300	4800	1600	3700
Total number of neurones	6000	10400	6300	16600
B. Relative proportions of different components (%)				
All neurones				
Sensory (all)	30	54	75	78
Sensory (unmyelinated)	18	31	62	54
Postganglionic	55	46	25	22
Neurones with unmyelinated axons				
Postganglionic	75	60	29	29
Postganglionic evaluated neurophysiologically	∿ 70		25 ± 5 (*n* = 6)	

Modified from *McLachlan* and *Jänig* (1983)

1. The highest number of labelled cells in each of the different experiments is a realistic estimate of the absolute number of neurones projecting into a given nerve.

2. No sympathetic postganglionic axons are myelinated (see Fig. 18).

3. Myelinated axons in hindlimb nerves do not branch between their cell body and the site of the nerve at which fibre counts were made several centimeters proximal to their termination. Branching of myelinated and unmyelinated axons in the sural and superficial peroneal nerve of the cat could not be demonstrated by electrophysiological tests (*Blumberg* and *Jänig*, unpublished observations).

4. The differences between the numbers of labelled sensory neurones in the dorsal root ganglia and the numbers of myelinated fibres in the respective peripheral nerves (taking additionally into account the numbers of skeletomotor axons in the muscle nerve) then lead to the number of sensory neurones with unmyelinated axons (*McLachlan* and *Jänig* 1983).

Table 4 shows the composition of four hindlimb nerves from which the postganglionic axons have been intensively investigated by neurophysiological methods (see also Table 1). Each nerve has a typical composition of nerve fibres. For example, the percentage of postganglionic axons, with respect to the total number of unmyelinated axons, is 75% in the muscle nerve (GS), 60% in the skin nerve to the paw (including hairy and hairless skin, MP) but only 29% in nerves to hairy skin (SU, SP). The proportions for the muscle nerve (GS) and for the nerves to the hairy skin (SU, SP) are very similar to the 70% and 25% ± 5%, respectively, sympathetic fibres identified electrophysiologically (*Blumberg* and *Jänig* 1982; *Jänig* and *Nelke,* unpublished observations). The high proportion of postganglionic axons in the muscle nerve and in the nerve to the paw presumably reflects the importance of the tissues which are innervated by these nerves in the regulation of peripheral vascular resistance and body temperature, respectively.

8 Effects of Anaesthetics

A continuing concern in neurophysiological analyses of discharge patterns of neurones in animals is the effects of the anaesthetics used. For those most commonly used in animal experiments (barbiturates, chloralose, ketamine, Althesin, inhalation anaesthetics), we have no sound knowledge of how and where they take effect. Consequently, we also have only limited knowledge about their effects on activity in different systems, not to mention specific reactions and reflexes. There is general agreement that most of them affect the cortex, limbic system and, to a certain degree, hypothalamic and brainstem functions and, therefore, also the functioning of sympathetic systems (*Gilman* et al. 1980). Also in general agreement is that α-glucochloralose, which is used worldwide in experimental studies on the autonomic nervous system in cats, dogs and rabbits, causes the least distortions of the functioning of the sympathetic systems. It may induce some excitation and therefore be responsible for the somewhat increased mean arterial BP in these animals.

In acute experiments on animals, ethical reasons necessitate the use of general anaesthesia. It is a fundamental error to believe that any distortion introduced by the anaesthesia can be avoided by, for example, decerebrating the animal; this procedure leads to changes in the central neuronal machinery and therefore to changes in the reaction patterns of the sympathetic systems (*Jänig* 1975; *Horeyseck* and *Jänig* 1973; *Korner* 1971, 1979). There is no doubt that different anaesthetics may affect the activity in a particular autonomic outflow in a different manner and that the effect of the same anaesthetic may vary depending on the autonomic outflow or reflex (*Gilman* et al. 1980; *Blake* and *Korner* 1982; *Blake* et al. 1982).

In order to get an approximate estimate of the effects of anaesthetics, these effects must be tested on ongoing activity and well-defined reflexes in functionally identified sympathetic systems. This is shown in summarized

Table 5. Effects of anaesthetics on SM activity and CVC activity. (Summarized from *Jänig* and *Räth* 1980)

	Methohexital (3–10 mg/kg i.m.)	Ketamine (14–23 mg/kg i.m)	Chloralose (4.5–20 mg/kg i.p.)	Althesin (2.5–5 mg/kg i.m.)	Halothane (0.5–1 vol. %)
Ongoing SM activity	↓[a]	↓[b]	↓[c]	↓ to 0	↓ to 0
Reflex on stimulation of Pacinian corpuscles (SM)	↓	↑	↑	↓	↓
Reflex on stimulation of cutaneous nociceptors (SM)	↑	↓	↑	↓	↓
Ongoing activity in CVC neurones	↓	=, ?	=, ?	↓ prob. to 0	↓ prob. to 0

↑, increase; ↓, decrease; =, no change

[a] Desynchronization of ongoing activity of SM activity between left and right hindpaw or forepaw

[b] Some desynchronization

[c] No or small effect on synchrony of ongoing SM activity; occasionally, large synchronous potentials

SM and CVC activity were determined indirectly by measuring skin potential and skin temperature on the surface of the central pad of the cat hindpaw respectively

form in Table 5 for ongoing activity in SM neurones and for two reflexes in the SM neurones, which are elicited by stimulation of Pacinian corpuscles in the paws and by stimulation of cutaneous nociceptors (by radiant heat or mechanically; see Sect. 3.3). All anaesthetics used depress ongoing activity, but each to a different degree; for example, halothane and Althesin depress it completely, the other anaesthetics, in moderate doses, to a much lesser extent. Methohexital desynchronizes the SM activity between both hindpaws or between hind- and forepaws; ketamine and α-chloralose have a milder desynchronizing effect on the SM activity. After chloralose, large completely synchronized discharges may appear during rest. Althesin and halothane depress SM reflexes, chloralose enhances both reflexes; ketamine enhances the "vibration" reflex in SM neurones and depresses the reflex caused by noxious stimulation of skin. Finally, methohexital enhances the reflex caused by noxious stimulation of skin and depresses the "vibration" reflex (*Jänig* and *Räth* 1980).

The above complex effects can only be obtained in brain-intact animals and not in chronic spinal animals. Thus, the anaesthetics probably affect the SM system at the brainstem and hypothalamic level or higher up, with the result that the activity in descending spinal systems, which control the spinal neuronal circuits, changes. The complexity of the effects of the anaesthetics indicates that each one influences the central neuronal machinery, which is involved in the regulation of sweat glands, in a different way.

Activity in CVC neurones (as judged by the skin temperature on the surface of the cat paw) is seriously depressed by Althesin and halothane, less by methohexital and probably even less by ketamine and α-chloralose (last row in Table 5). These results agree with those of *Wallin* and *König* (1976), who showed that skin nerve sympathetic activity is depressed during anaesthesia with thiopentone in humans.

9 Comparison of the Discharge Patterns of Postganglionic Neurones Supplying Skeletal Muscle and Skin in Cats with Those in Humans

The introduction of microneurography has made it possible to study multi- and single-unit axon activity in peripheral nerves of the human being; this technique enables the study of activity in sympathetic postganglionic axons in human muscle and skin nerves in non-anaesthetized subjects who can freely communicate, independent of the effector responses (for example, blood flow through skeletal muscle and skin, sweat secretion), and allows correlation of these neural activities with the effector responses (*Valbo* et al. 1979). This section summarizes the similarities and differences, between humans and cats, in the discharge patterns of postganglionic neurones supplying skeletal muscle and skin (*Jänig* et al. 1983). Initially, some advantages and limitations of animal experiments versus recordings in humans are discussed.

9.1 Animal Experiments Versus Experiments on Humans

Experiments on humans may serve to test the validity of hypotheses which have been developed from animal experiments or vice versa. Yet both forms of recording nerve activity have their advantages and limitations for analyzing the functioning and organization of the sympathetic nervous system. The interaction between both approaches will bring out the real potentialities of either method.

A main advantage of microneurography is that one can work with unanaesthetized human subjects in whom all higher brain structures are functioning. Activity in sympathetic postganglionic neurone populations can be studied during ongoing regulations of effector organs without interference by major surgery. Relatively safe, functional identification of postganglionic neurones is possible at least for the MVC, CVC and SM neurones, by way of the discharge patterns and the responses to various manoeuvres and by correlating the postganglionic activity with effector responses (e.g. blood flow through skin and skeletal muscle, skin resistance changes). Furthermore, microneurography can and will be a valuable tool in the analysis of pathophysiological states of the sympathetic nervous system. The latter application may not only deepen the insight in the neuronal mechanisms of dysregulation of sympathetic activity, but may in addition yield information on the principles of the function of the sympathetic nervous system in health. The human patient may serve, in this respect, as a "natural" model.

There are also some limitations to the application of microneurography, which must be kept in mind. Recording of postganglionic activity in human nerves only rarely allows a single-unit analysis (*Hallin* and *Torebjörk* 1974). Furthermore, up to now there has been no successful analysis of postganglionic neurones which are normally silent and can only be activated under very specific functional conditions. This applies particularly to MVD, CVD and PM neurones. It is unclear whether postganglionic MVD and CVD neurones supplying blood vessels of the skeletal muscle and skin, respectively, exist at all in humans (but see *Rowell* 1981). In addition, the postganglionic supply of functionally different skin areas, such as hairless skin of the hands, hairy skin of the forearm and skin of the face, varies in its composition (*Bini* et al. 1980b). Thus, it may well be that the sympathetic supply to the face lacks postganglionic CVC axons but that the sympathetic supply to the hairless skin does not.

The experimental situation in microneurography on humans is much more restricted than the situation in animal experiments. Ethical and medical standards do not allow the use of major invasive techniques if not required for diagnostic and therapeutic purposes. In non-anaesthetized human beings, changes of attention, of state of vigilance, alerting situations, emotional situations, etc. may at once be reflected in changes of activity in postganglionic CVC and SM neurones. This certainly interferes with the experimental protocol and may therefore mask responses which can be elicited in cats (for example, when a noxious stimulus is applied to the skin, see below). Thus, the unanaesthetized awake state not only has advantages but also some limitations which cannot be fully controlled and which are difficult to overcome. In the cat, for example, stimulation of Pacinian corpuscles in the hindpaw leads to excitation of SM neurones supplying sweat glands in the hairless skin of the hindpaw. This is probably a spinal reflex (see Sects. 3.3, 4.4). This reflex cannot be obtained in SM neurones in unanaesthetized humans. However, under certain pathophysiological conditions (e.g. hyperhidrosis, see below, Sect. 9.4), it can be shown to exist in humans. This finding may mean that the spinal reflex pathway, which enables the "SM vibration reflex", is normally masked in humans (e.g. by descending inhibition) and that it is used only under very special functional conditions. It does not necessarily mean that this spinal pathway is normally not working and unimportant in humans. Theoretically, the activity from Pacinian corpuscles in the distal extremities could be a positive feedback signal in order to facilitate the activation of SM neurones from supraspinal brain structures.

Advantages and limitations of experiments on cats are nearly complementary to those on human beings. Ethical considerations necessitate the use of general anaesthesia. This raises the question about the distortion of the reflex patterns (see Sect. 8) and to what extent the results are applicable to awake animals. Also, the surgery may contribute to these distortions (e.g. by producing large nociceptive inputs, see Sects. 4.3, 4.4). The obvious advantage of animal experiments is that controlled experiments on single pre- and postganglionic units and on neurones in the central nervous system can be performed under stable conditions. However, to what extent results obtained in this way and hypotheses developed from these experiments are relevant for the neuronal processes during ongoing regulations of sympathetic target organs can be tested only in unrestrained awake animals and human beings.

9.2 MVC Neurones

Postganglionic MVC neurones in recumbent humans have ongoing activity. The rate of ongoing activity in single postganglionic MVC neurones is unknown, however, it would be very surprising if it were much different from that found in the cat (probably 1–2 imp/s or lower). By counting the multi-unit bursts of activity per 100 heart beats (or per time period), it was found that the ongoing discharge ranged from less than 10 to 90 bursts per 100 heart beats in different individuals. The level of activity remained fairly constant for a given subject over several months (*Sundlöf* and *Wallin* 1977).

The ongoing activity in MVC neurones of humans displays phenomenologically the same pulsatile modulation (cardiac rhythmicity; see Sect. 3.1) and respiratory modulation ("respiratory rhythmicity"; see Sect. 3.1) as in the cat (*Delius* et al. 1972a,b; *Sundlöf* and *Wallin* 1978; *Bini* et al. 1981). The pulsatile modulation of activity is most likely linked to the pulsatile excitation of arterial baroreceptors by the pulse pressure wave. The respiratory modulation of MVC activity is closely linked to respiratory fluctuations of the arterial BP: the discharges occur with decreasing BP, i.e. with unloading of arterial baroreceptors. Owing to this close linkage, it is mostly impossible to elucidate in humans whether there is a close central coupling of respiratory neurones and neurones involved in the regulation of pre-postganglionic activity to resistance vessels. The effects of various manoeuvres or stimuli on activity in MVC neurones in humans is mostly dominated (or masked) by the influence of changes of activity in arterial baroreceptors. Any procedure (e.g. orthostatic stress, raising of the legs, Valsalva's manoeuvre, deep breathing, arousal, emotional stress) which leads to a change of arterial BP, and therefore to an increase or decrease of activity in arterial baroreceptor afferents, is followed by the expected changes of MVC activity (i.e. decrease or increase; see *Wallin* 1981). Static changes of arterial BP seem to have only a transient influence on MVC neurone activity in humans. Also, continuous activation of arterial baroreceptor afferents induces only a transient decrease of postganglionic activity to skeletal muscle (*Wallin* et al. 1975; *Båth* et al. 1981; *Wallin* and *Eckberg* 1982). These findings in humans are fully consistent with the results obtained on MVC neurones in cats (see Sect. 3.1.2).

In human beings, application of maintained lower body negative pressure is followed by maintained increase of activity in MVC neurones, without statistically significant changes of arterial BP (*Sundlöf* and *Wallin* 1978). This procedure induces displacement of the central blood volume from the thoracic cavity to the abdominal cavity and to the lower extremities and consequently to an unloading of intrathoracic volume receptors (*Wolthuis* et al. 1974). Therefore, it is believed that the maintained increase of MVC activity is induced by static unloading of these volume receptors.

Generally, it can be stated that MVC neurones, in unanaesthetized humans and anaesthetized cats, have very similar discharge characteristics.

9.3 CVC Neurones

In humans, CVC neurones have ongoing activity. The rate of ongoing activity and the proportion of CVC neurones with ongoing activity are dependent on the ambient temperature and mental state of the subject. Furthermore, there seem to be considerable differences between the CVC activity to the distal skin areas and that to the proximal skin areas of the extremities; at normal room temperature, CVC neurones to distal skin areas have ongoing activity, whereas most CVC neurones to proximal skin areas seem to be silent (*Bini* et al. 1980b). The ongoing activity in CVC neurones is not pulse-synchronous and remains unchanged when a carotid sinus nerve is stimulated electrically (*Hagbarth* et al. 1972; *Wallin* et al. 1975); thus, the activity in arterial baroreceptor afferents seems to have no influence on the CVC system in humans. There is a loose and variable coupling of respiration and activity in CVC neurones (*Hagbarth* et al. 1972).

Thermal stimuli are the most specific stimuli which change the activity in CVC neurones supplying distal skin areas: exposure to warm or cold environments is followed by a decrease or increase in CVC activity, respectively (*Bini* et al. 1980a). Otherwise, arousal or mental (emotional) stimuli, and also deep breaths, hyperventilation and muscle work, may activate CVC neurones (*Delius* et al. 1972c; *Hallin* and *Torebjörk* 1974; *Wallin* 1981). This latter type of activation of CVC neurones is typical for the unanaesthetized subject and probably dependent on the functioning of the cerebral cortex and limbic system. It may mask more detailed reactions of CVC neurones to somatic and visceral stimuli which can be seen in anaesthetized animals.

Recently *Blumberg* and *Wallin* (personal communication) succeeded to the elicit reflex vasodilation (increase of blood flow) in human skin of the feet by electrical intraneural stimulation of afferent nerve fibres. This dilation was abolished by blockade of the conduction in the nerve proximal to the stimulation site; it was largest on the stimulated hindlimb and smaller on the contralateral hindlimb. The blood flow through the skin of the hand did either decreased or did not change. The dilation was particularly prominent when the subject's skin was cooled, i.e. when the activity in the CVC neurones was high. Thus, the authors measured a reflex which had a spatial organization and it is very likely that it consisted of inhibition of CVC activity. This reflex in humans could be identical to the inhibitory reflex in CVC neurones elicited by cutaneous noxious stimuli in cats (see Figs. 5E, 6, 11B,C) and to the Lovén reflex (*Lovén* 1866).

As judged by its discharge pattern, it may appear that the CVC system in humans is considerably different from that in cats (see Sect. 3.2). However, it must be kept in mind that the pattern of CVC activity in the cat has been obtained under general anaesthesia. In this condition, one may see the fine details of the discharge patterns which are dependent on the neuronal circuits in spinal cord and brainstem and which can barely be seen in the unanaesthetized state.

9.4 SM Neurones

SM neurones in humans are active at high ambient temperatures and silent at low ambient temperatures. Otherwise, they are activated by arousal and mental (emotional) stimuli in the same way as CVC neurones (*Bini* et al. 1980a; *Wallin* 1981). SM neurones supplying more proximal hairy skin areas have a lower threshold in their firing to warming of the body surface than SM neurones supplying hairless skin. These latter SM neurones are only activated at relatively high ambient temperatures which are unpleasant

(*Bini* et al. 1980b). Surprisingly, the postganglionic SM activity at these temperatures shows cardiac rhythmicity, as if the SM system in humans were under control of the arterial baroreceptors (*Bini* et al. 1981).

In the cat, SM neurones are probably not involved in thermoregulation; they may be functionally homologous to the SM neurones supplying hairless skin in humans. In both species, one important function of this SM supply may be in keeping the hairless skin flexible for optimal sensory discrimination. In this context, it is interesting to note that SM neurones supplying glabrous skin of the hands of hyperhidrotic patients (who suffer from uninhibited profuse sweating) can be activated by stimulation of low-threshold mechanoreceptors (probably Pacinian corpuscles of the hand; *Marchettini, Torebjörk, Culp, Ochoa,* personal communication). This reflex is similar to the vibration reflex in the SM neurones of the cat and probably due to a spinal pathway. Observations of hyperhidrotic patients show that the reflex pathway which mediates the "vibration reflex" to the SM neurones also exists in humans. This reflex pathway normally seems to be suppressed; alternatively, it cannot be excluded that it is also viable in healthy humans during certain functional conditions.

9.5 Reciprocal Changes of Blood Flow Through Skeletal Muscle and Skin in Humans

Resting blood flow through skeletal muscle and skin of limbs in relaxed humans exhibits continuous fluctuations under thermal neutrality. The intervals between the fluctuations are about 0.5–2 min. The blood flow changes are synchronous in the four extremities (Fig. 19A, upper two traces; *Golenhofen* 1971).

The spontaneous changes of blood flow through skeletal muscle are reciprocal to those through skin (Fig. 19A). Peak values in muscle blood flow coincide with minimal values in skin blood flow and vice versa. Mild and strong emotional stimuli also induce reciprocal changes in blood through both vascular beds (Fig. 19B). Skin blood flow decreases and muscle blood flow increases (*Golenhofen* and *Hildebrandt* 1957; *Golenhofen* et al. 1961; *Golenhofen* 1971).

The spontaneous fluctuations and evoked changes of blood flow are dependent on the activity in postganglionic sympathetic neurones supplying the vascular beds. Blockade of the conduction in the respective nerves abolishes the blood flow changes nearly completely. Adrenaline released from the adrenal medulla may have some weak residual reciprocal effect on both vascular beds (*Golenhofen* et al. 1961). It is unclear whether the increase of blood flow through skeletal muscle during emotional stimuli is produced by decrease of activity in MVC neurones or by activation of

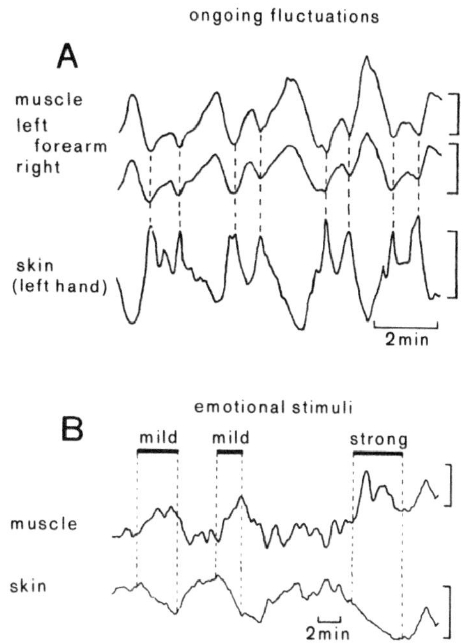

Fig. 19A,B. Spontaneous and evoked changes of blood flow through skin and skeletal muscle of human limbs. The blood flow through skeletal muscle is determined by heat conduction probes, those through skin by flow calorimeters. In **B** the blood flow through muscle was determined at the left calf and through skin at the sole of the right foot. The mild emotional stimuli in **B** were produced by reading descriptions of fights in a book, the strong emotional stimulus by mental arithmetic. The *ordinate scales* indicate 50% or 100% (muscle in **B**) of the mean blood flow. (*Golenhofen* and *Hildebrandt* 1957; *Golenhofen* 1971)

vasodilator neurones. Atropine (applied intra-arterially to the extremity concerned) is without effect (but see *Blair* et al. 1959). Phentolamine infused intra-arterially leads to an increase of blood flow by about 100%, but does not abolish the vasodilator response during emotional stimuli (*Golenhofen* et al. 1961). It could well be that the dosage of phentolamine which was infused intra-arterially (0.8 mg/min) was too low in order to block the neural α-adrenergic effect on the muscle vascular bed completely.

In humans, postganglionic MVC neurones supplying different muscles of the hind- and forelimbs discharge synchronously; this also applies to CVC neurones (*Sundlöf* and *Wallin* 1977; *Bini* et al. 1980b). However, ongoing long-term fluctuations of sympathetic activity in cycles of 0.5–2 min and reciprocity between ongoing discharges in MVC and CVC neurones have not been described in the literature. Emotional stimulis induce excition of CVC neurones and sometimes weak depression of MVC neurones (*Vallbo* et al. 1979; *Wallin* 1981). Reciprocal spontaneous changes of activity in MVC and CVC neurones can occasionally also be obtained in anaesthetized, immobilized and ventilated cats in which the carotid sinus and vagoaortic nerves are cut (*Jänig* 1979).

It may be hypothesized that spontaneous and evoked reciprocal changes of blood flow through skeletal muscle and skin are produced by reciprocal changes of activity in MVC and CVC neurones. The reciprocity may be generated by neuronal circuits in the spinal cord ("spinal sympathetic functional units", see Sect. 10) which are reciprocally organized (see Fig. 12).

10 General Conclusions and Future Perspectives

This review has focussed on *two complementary aspects* of the sympathetic outflow to skeletal muscle and skin of the cat hindlimb and tail. First, the existence of several separate sympathetic pre-postganglionic channels to different autonomic target organs, or even to the same autonomic target organ (e.g. blood vessels; Fig. 1), has been discussed. The main arguments that the pre-postganglionic sympathetic pathways are separate derive from the characteristic discharge patterns of the pre-postganglionic neurones and from the different distributions of the conduction velocities of the pre- and postganglionic axons for each functional type of neurone. The problem of transmission and integration of impulses in the paravertebral ganglia, with respect to the different types of postganglionic neurones, will be discussed elsewhere. Neither the characteristic discharge patterns of the different types of neurones nor the characteristic distributions of the conduction velocities of their axons prove that the channels to the target organs are really separate. It is possible that there is considerable overlap in the pre-postganglionic synaptic connections between different systems; this would, however, imply that only certain synaptic connections normally lead to suprathreshold excitations of postganglionic neurones and that others do not. How else can one explain the characteristic reaction pattern of each type of postganglionic neurone. The possible overlap in the pre-postganglionic synaptic connections between different systems can only be investigated with intracellular measurements in postganglionic neurones in vivo when the preganglionic axons are intact (see *Blackman* 1974; *Skok* 1973; *Skok* and *Ivanov* 1983). The implication of the distinct discharge patterns is that pre- and postganglionic neurones can be functionally identified independent of their target organs in a given, well-defined experimental situation (see Sect. 5.2).

Second, the discharge patterns of different types of sympathetic neurones are dependent on the structure of the respective system in the neuraxis. We do not know the details of these neuronal structures. However, the global organization of different sympathetic systems in the periphery and in the neuraxis (spinal cord and brainstem) can be described. Thus, the reaction (reflex) patterns in the sympathetic systems may serve as leads for speculations about the central organization of the systems; in other words, we can look into the neuronal structures of the neuraxis via these discharge patterns (*Jänig* 1975, 1979, 1982).

10.1 Reflexes as Tools

The main arguments used in this review are derived from measurements of reflexes in post- and preganglionic neurones. These reflexes were elicited by adequate stimulation of receptor populations from the body surface and from its interior. The application of the *reflex concept* to the sympathetic nervous system may appear old-fashioned and too narrow a concept for elucidating the neural regulations of autonomic effector organs by the sympathetic nervous system. This is certainly true. Reflexes, as described above, are mostly isolated fragments of regulating systems and, as such, experimental artefacts. Some reflexes may easily be interpreted as functionally meaningful (e.g. vibration reflex, baroreceptor reflex, chemoreceptor reflex), while others may not. However, the application of the reflex concept provides an insight into the neuronal elements of the regulation of sympathetic target organs. The advantages of this approach are obvious (*Granit* 1981): (a) one knows the types of efferent neurones being controlled, (b) the afferent neurones stimulated are known, (c) the experimental conditions can be reasonably well defined and (d) the reflexes can be studied in various types of preparations (brain-intact, decerebrate, spinal animals; animals with other central lesions). Once the reflexes and their pathways (including interneurones, etc.) are well defined, one should be able to elucidate how these pathways function and interact during ongoing regulations of autonomic target organs.

Two reflexes which are functionally reasonably well understood may serve as examples. The arterial baroreceptor reflex is perhaps the best known element of nervous control of autonomic cardiovascular function. For decades it served as a paradigm for elucidating the neuronal control of vascular resistance and cardiac output, analogous to the role of the monosynaptic stretch reflex in research on neural regulation of movement (*Homma* 1976). The discharge properties of the afferent input of the baroreceptor reflex can be well described (*Kirchheim* 1976); the integration of this afferent activity by neurones in the nucleus tractus solitarii (*Spyer* 1981) and the efferent postganglionic output to resistance vessels and to the heart can be reasonably well described. However, our knowledge about the pathways and integration in the medulla oblongata, about the descending spinal pathways and about the integration in the spinal cord by interneurones and preganglionic neurones is fragmentary. The synaptic transmitters and their membrane effects are almost completely unknown (*Spyer* 1981; *Abboud* and *Thames* 1983). It is not farfetched to predict that the arterial baroreceptor reflex will further serve as a basic neuronal unit in research on the neural regulation of vascular resistance and cardiac output for the next 10–20 years. The second example is the "vibration reflex" in the SM neurones (see Sects. 3.3, 4.4). This specific reflex is spinal. The discharge patterns of the afferent neurones and the postganglionic SM neurones can be well described. The elements of the spinal reflex are and its control by spinal descending systems is only marginally known (*Jänig* and *Kümmel* 1981).

Frequently, it cannot be predicted how the ensemble of neurones, being interposed in different reflex pathways, functions during ongoing regulations since the systems may become too complex, but one may be able to develop hypotheses which are testable in animals and humans under closed-loop conditions. The situation is conceptually and methodologically analogous to the development of analyses of the neuronal regulation of the skeletal muscle during movement. This field initially obtained its impetus from the application of the reflex concept and the subsequent formulation of hypotheses which could be tested during ongoing movements in animals and humans (*Granit* 1981; *Baldissera* et al. 1981).

10.2 Hierarchical Organization of Lumbar Sympathetic Systems and Spinal Sympathetic Functional Units

A general idea about the organization of the sympathetic systems in the periphery and neuraxis is schematically exemplified in Fig. 20. This scheme includes several ideas which have been borrowed from the neurophysiological field of the motor system. Though the scheme is speculative, it may help to guide future neurobiological analyses of the sympathetic nervous system in somewhat different directions. It may furthermore help to reconcile ideas about the general functioning of the sympathetic nervous system and newly developing ideas about the specificity of different sympathetic systems.

The sympathetic systems have a hierarchical organization. At the bottom of the organization are the functionally separated pre-postganglionic channels. This (functional) separation does not preclude a complex integration of neuronal activity in the sympathetic ganglia. The degree of convergence and divergence of preganglionic axons for different sympathetic systems is unknown. However, it is known that synaptic cholinergic muscarinic and non-cholinergic long-term processes are built in the vasoconstrictor pathways but probably not in the other pathways (*Hoffmeister* et al. 1978; *Jänig* et al. 1982, 1983; *Blumberg* and *Jänig* 1983). The size of the neural unit in these pathways, which is the equivalent of the motor unit, is unknown (see *Purves* and *Wigston* 1983). It would not be surprising if this neural unit varies between different sympathetic channels.

The investigation of the reflex patterns of different types of sympathetic neurones, in brain-intact and chronic spinal animals, indicates that the

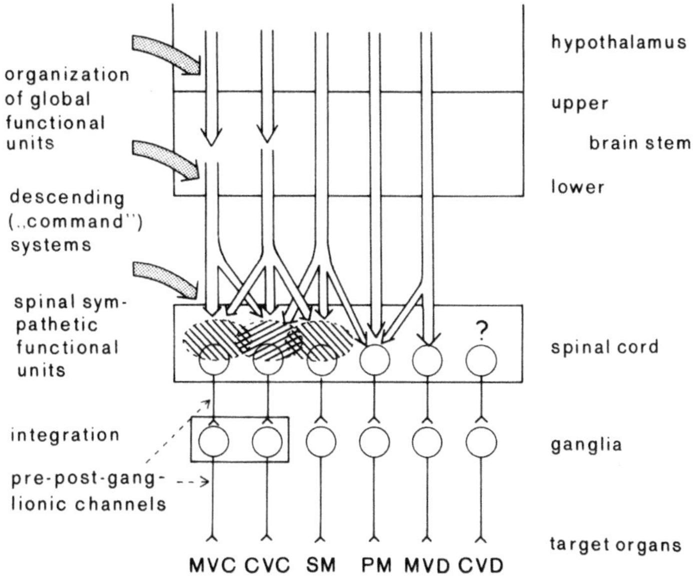

Fig. 20. General ideas and speculations about the organization of the sympathetic systems. Note not only the vertical but also the horizontal organizations, particularly at the level of the spinal cord. The ***shaded arrows*** on the ***left*** side symbolize the afferent inputs (e.g. primary afferents from visceral organs and somatic structures, afferents from arterial baro- and chemoreceptors, thermoreceptive inputs from spinal canal and hypothalamus, input from the ventral side of the medulla oblongata). The problem of transmission of impulses in sympathetic ganglia is not treated in this article (see *Hoffmeister* et al. 1978; *Blumberg* and *Jänig* 1983a; *Jänig* et al. 1982, 1983). (Modified from *Jänig* 1984)

spinal cord probably is important for the generation of these patterns (Fig. 12). For each sympathetic system, a spinal neuronal circuit may exist. This appears quite obvious for the SM and the CVC systems (*Jänig* and *Kümmel* 1981). A circuit may consist of preganglionic neurones, interneurones and its synaptic connections with the primary afferent input and descending spinal systems. This neuronal ensemble may be called *"spinal sympathetic functional unit"*. Related spinal sympathetic functional units may share common interneurone pools, primary afferent inputs and descending spinal systems from the brainstem and hypothalamus.

The spinal sympathetic functional units receive information from the periphery by primary afferent neurones and from the brainstem and hypothalamus by descending spinal pathways. Both sets of inputs converge probably on common interneurones; the descending spinal inputs may additionally converge directly on preganglionic neurones (*Spyer* 1981). Conclusive evidence for the latter is missing. Primary afferent neurones probably do not synapse monosynaptically with preganglionic neurones (see *Sato* and *Schmidt* 1973; *Coote* 1978).

Both, primary afferent inputs and descending inputs must have certain selectivity with respect to the spinal sympathetic functional units. The specificity of the reaction patterns of preganglionic neurones does not mean that preganglionic neurones (and interneurones) of a particular spinal sympathetic functional unit receive synaptic input selectively from certain primary afferents and spinal descending pathways and not from others. It could well be that there is a general convergence of primary afferents and descending axons on these neurones and that normally only certain synaptic inputs are dominant and reach the threshold, but others not or only in very unusual conditions. Furthermore, it is likely that part of the specificity of the reaction patterns is encoded in the neuronal structures of the spinal sympathetic functional units.

From the functional point of view, several spinal descending paths from the brainstem and hypothalamus to the spinal sympathetic functional units must be postulated; otherwise, the complexity of the responses of pre- and postganglionic neurones to various peripheral and central stimuli can barely be understood. Fluorescence microscopic and immunohistochemical investigations and studies with HRP show that there are several spatially and chemically distinct spinal descending projections to the intermediolateral cell column. These projections originate from circumscribed neurone groups in the lower brainstem, upper brainstem and hypothalamus (*Amendt* et al. 1979; *Loewy* and *Neil* 1981; *Loewy* 1982). They may release noradrenaline, adrenaline, serotonin and oxytocin as transmitters but probably also other substances (*Loewy* and *Neil* 1981; *Loewy* 1982). The influence of spinal descending pathways on the thoracolumbar sympathetic outflow was investigated physiologically using recordings of BP and sympathetic nerve activity (renal nerve, splanchnic nerve, WR). In these studies, distinct sites in the spinal cord and lower brainstem were electrically stimulated and lesioned. The studies show that sympathetic preganglionic neurones can be excited and inhibited from the brainstem via various descending pathways running in the dorsolateral funiculus and in the ventral spinal cord. Excitation can be elicited from the dorsolateral funiculus (*Illert* and *Gabriel* 1972; *Foreman* and *Wurster* 1973; *Coote* and *MacLeod* 1974a,b); the dorsolateral descending pathways have a viscerotropic organization (*Barman* and *Wurster* 1975). Inhibition can be elicited from the dorsolateral funiculus as well as from the ventrolateral and ventral funiculus (*Illert* and *Seller* 1969; *Coote* and *MacLeod* 1974a; *Dembrowsky* et al. 1981). Indirect evidence exists that baro- and chemoreceptor reflexes to preganglionic neurones with functional properties of MVC neurones may be relayed by separate spinal descending pathways (*Jänig* and *Szulczyk* 1979), that the SM system is controlled by two descending pathways (*Jänig* and *Kümmel* 1981) and that the hypothalamic control of CVC neurones (thermoregulation!) is different from the hypothalamic control of the MVD system. Critical reviews on the functional organization of spinal descending pathways possibly influencing cardiovascular preganglionic neurones are given by *Dampney* (1981) and *Spyer* (1981).

The descending systems from the brainstem and hypothalamus use the spinal sympathetic functional units as basic sympathetic motor subroutines. The relation between the descending systems and the spinal sympathetic functional units varies according to the function of these units. For example, for the descending paths relaying information from arterial baro- and chemoreceptors to the MVC system, this relation may probably be more direct and dominant. The relation between these descending systems and the CVC system may be more indirect. Furthermore, the MVC system may be more under the control of the lower brainstem (*Gregor* and *Jänig* 1977) and the MVD, SM, PM and CVD systems more under the control of the upper part of the neuraxis. The CVC system may be normally under the control of the rostral neuraxis.

In summary, it may be hypothesized that the spinal cord is much more important for the regulation of sympathetic target organs than hitherto assumed. It contains the spinal functional units. These units, rather than their individual components (such as the preganglionic neurones) are operated during reflexes initiated by activation of spinal afferents and by central programmes of autonomic regulation via the spinal descending pathways. This spinal organization of the sympathetic nervous system would be similar to the neuronal organization of spinal reflex arcs for regulation of movements by skeletal muscles (*Granit* 1975; *Lundberg* 1975, 1979a,b; *Baldissera* et al. 1981).

We must not be content to assume that the spinal sympathetic functional units and their synaptic connections with the primary afferent inflow and with the descending systems from the brainstem and hypothalamus function always in the same manner as a wiring diagram may imply. The pattern of the operation of the functional units may change in different functional states of the organism; therefore, also the discharge patterns in the pre- and postganglionic neurones may change. For example, most CVC neurones exhibit a typical discharge pattern under our experimental conditions in brain-intact anaesthetized cats (Fig. 12). After decerebration of the animals, the inhibitory chemoreceptor reflex in these CVC neurones is reversed to an excitatory reflex (*Gregor* and *Jänig* 1977). Rabbits show similar changes in sympathetic activity to skin and in blood flow, in response to stimulation of arterial chemoreceptors after decerebration (*Iriki* and *Kozawa* 1976). These results may be interpreted to mean that the CVC system is not only under control of the rostral neuraxis but also under control of the caudal neuraxis. The latter control may be similar to that of the MVC system; it is usually masked and therefore not seen in most CVC neurones. However, this control may appear in special conditions.

One such situation is possibly produced by a chronic lesion of a skin nerve at the hindlimb (e.g. by cutting and ligating the superficial peroneal nerve or by connecting the central stump of the superficial peroneal nerve to peripheral stumps of the deep peroneal nerve; see *Blumberg* and *Jänig* 1983b, 1985). In this condition many CVC neurones, in the lesioned nerve as well as in a contiguous skin nerve, have a discharge pattern which is similar to that in MVC neurones. The reciprocity seen in the discharges between MVC neurones and CVC neurones is then partially or completely lost (for details see *Blumberg* and *Jänig* 1983b, 1985).

10.3 Questions

Future research will have to approach this system from the vantage point of the functionally identified preganglionic neurone at the spinal level. Experimental (neurophysiological, morphological, histochemical and neuropharmacological) work should concentrate on the following questions:

1. Are functionally different types of preganglionic neurones situated at different sites in the spinal cord? Do they differ in the orientation of their dendrites?
2. What is the composition of a spinal sympathetic functional unit? Which are the functional characteristics and locations of autonomic spinal interneurones?
3. Can one discriminate between different types of descending spinal pathways with respect to functionally identified preganglionic neurones.
4. In which way do descending spinal pathways gain access to spinal sympathetic functional units: mainly via interneurones or also directly to preganglionic neurones?
5. How important is the input from the periphery via spinal primary afferent neurones for the functioning of the spinal sympathetic functional units?

Reasonable answers to these questions may lead to changes in our way of thinking about the functioning and organization of the sympathetic systems. More emphasis will be put on the spinal cord. The diversity of reflexes and reactions, which can be measured in different types of post- and preganglionic neurones, may then be explained and unified in new concepts about the organization of the sympathetic systems.

11 Summary

The lumbar sympathetic outflow supplying the skeletal muscle and skin of the hindlimb and tail of the cat is composed of six pre-postganglionic channels: muscle vasoconstrictor (MVC), cutaneous vasoconstrictor (CVC), sudomotor (SM), pilomotor (PM), muscle vasodilator (MVD) and cutaneous vasodilator (CVC) neurones.

1. Most postganglionic MVC, CVC and SM neurones have ongoing activity. Functionally, they are characterized by way of the reflex (reaction) patterns produced by adequate stimulation of receptor populations on the surface and in the interior of the body. *MVC neurones* are under inhibitory control of arterial baroreceptors; most other stimuli lead to excitation of these neurones. Most *CVC neurones* are under inhibitory control of most afferent inputs; arterial baroreceptors have only weak effects. Some CVC neurones behave like MVC neurones. *SM neurones* are excited by all afferent stimuli; stimulation of Pacinian corpuscles in the paws (by vibration) leads to selective activation of SM neurones. These reflex patterns also exist at the preganglionic level.
2. Postganglionic PM, MVD and CVD neurones are silent. These neurones can only be activated by specific central stimuli. The existence of CVD neurones is not as well established as that of the other types of postganglionic neurones.
3. Most reflexes elicited in MVC, CVC and SM neurones by stimulation of cutaneous and visceral receptors and spinal thermoreceptors are preserved in *chronic thoracic spinal cats.* The only major change which occurs is that visceral stimuli, which produce inhibition in CVC neurones in animals with an intact neuraxis, elicit excitation of these neurones in spinal animals. Furthermore, noxious stimulation of the skin of the ipsilateral hindpaw may elicit inhibition of ongoing activity in CVC neurones that lasts longer in spinal animals than in brain-intact animals and long-lasting afterdischarges in SM neurones.
4. Pre- and postganglionic neurones supplying the hindlimb and tail have *distinct segmental locations.* Most postganglionic cell bodies supplying the skin and skeletal muscle of the hindlimb are situated in ganglia L_5 to L_7. Of the postganglionic and afferent cell bodies which project into a hindlimb nerve, 90% or more lie in corresponding sympathetic and dorsal root ganglia. More than 95% of the preganglionic cell bodies which project into postganglionic neurones supplying the hindlimb and tail lie in the spinal segments T_{13} to L_4. The physiology and morphology of the segmental distributions of pre- and postganglionic neurones correspond closely.

5. Axons of pre- and postganglionic MVC, CVC, SM and PM neurones have characteristic distributions of *conduction velocities.* Conduction velocities of most *preganglionic* axons range from about 0.5 to 15 m/s. Preganglionic PM and SM axons are the fastest conducting ones and MVC axons the slowest. Conduction velocities of preganglionic CVC axons are intermediate. The conduction velocities of most *postganglionic* axons range from 0.3 to 1.2 m/s. All of them are unmyelinated. CVC axons are the slowest conducting ones (mean about 0.6 m/s); MVC, SM and PM axons conduct faster (means about 0.8 m/s).

6. Reflex (reaction) patterns, segmental distributions and dimensional characteristics of pre- and postganglionic neurones indicate that the final pre-postganglionic sympathetic paths to the target organs are, at least functionally, separate. The reflex patterns observed in chronic spinal animals, in comparison to those observed in animals with an intact neuraxis, indicate that the important neuronal circuits which determine these patterns probably lie in the spinal cord. The hypothetical spinal neuronal circuits which produce the reflex patterns are tentatively called *"spinal sympathetic functional units".* On the one hand, the reflex patterns of functionally different types of pre- and postganglionic neurones may serve as functional labels for recognizing these neurones; on the other, they may serve as a conceptual framework for the cellular analysis of the neuronal circuits in the neuraxis, which are the basis for the characteristic functional features of different types of pre- and postganglionic neurones.

Acknowledgements. This work was supported by the Deutsche Forschungsgemeinschaft. I would like to thank Dr. Klaus Dembowsky, Dr. Anders Lundberg and Dr. Gunnar Wallin for their criticism of the manuscript and their valuable suggestions. I also would like to express my gratitude to Dr. Elspeth McLachlan, who participated in continuous (valuable) discussions without once seeing the text. Last but not least, I would like to thank Dr. Paul Korner for letting me relax in the library of the Baker Medical Research Institute. The repeated typing of the text by Brigitte von Wieding and Helena Dawson is also appreciated.

References

Abboud FM, Thames MD (1983) Interaction of cardiovascular reflexes in circulatory control. In: Shepherd JT, Abboud FM (eds) The cardiovascular system. Vol III Peripheral circulation and organ blood flow. Part 2 Handbook of physiology. American Physiological Society, Bethesda, pp 675–753

Abboud FM, Heistad DD, Mark AL, Schmid PG (1975) Differential responses of the coronary circulation and other vascular beds to chemoreceptor stimulation. In: Purves MJ (ed) The peripheral arterial chemoreceptors. Cambridge University Press, London, pp 427–442

Abboud FM, Heistad DD, Mark AL, Schmid PG (1976) Reflex control of the peripheral circulation. Prog Cardiovasc Dis 18:371–403

Abrahams VC, Hilton SM, Zbrozyna AW (1960) Active muscle vasodilatation produced by stimulation of the brain stem: its significance in the defence reaction. J Physiol (Lond) 154:491–513

Abrahams VC, Hilton SM, Zbrozyna AW (1964) The role of active muscle vasodilatation in the alerting stage of the defence reaction. J Physiol (Lond) 171:189–202

Akoev GN (1981) Catecholamines, acetylcholine and excitability of mechanoreceptors. Prog Neurobiol 15:269–294

Amendt K, Czacharski J, Dembowsky K, Seller H (1979) Bulbospinal projections to the intermediolateral cell column; a neuroanatomical study. J Auton Nerv Syst 1: 103–117

Andersson PO (1983) Comparative vascular effects of stimulation continuously and in bursts of the sympathetic nerves to cat skeletal muscle. Acta Physiol Scand 118: 343–348

Baldissera F, Hultborn H, Illert M (1981) Integration in spinal neuronal systems. In: Brooks VB (ed) Motor control Part I. American Physiological Society, Bethesda, pp 509–595 (Handbook of physiology, section 1, The nervous system, vol II)

Barasi S, Lynn B (1983) Effects of sympathetic stimulation on mechanoreceptor and nociceptor afferent units with small myelinated (A-δ) and unmyelinated (C) axons innervating the rabbit pinna. J Physiol (Lond) 341:41P

Bard P (1928) A diencephalic mechanism for the expression of rage with special reference to the sympathetic nervous system. Am J Physiol 84:490–515

Bard P, Macht MB (1958) The behaviour of chronically decerebrated cats. In: Neurological basis of behaviour, Ciba Foundation Symposium. Churchill, London, pp 55–75

Bard P, Rioch DMcK (1937) A study of four cats deprived of neocortex and additional portions of the forebrain. Bull Johns Hopkins Hosp 60:73–147

Barker D, Saito M (1981) Autonomic innervation of receptors and muscle fibres in cat skeletal muscle. Proc R Soc Lond (Biol) 212:317–332

Barman SM, Wurster RD (1975) Visceromotor organization within descending spinal sympathetic pathways in the dog. Circ Res 37:209–214

Baron R, Jänig W, McLachlan E (1985a) On the anatomical organization of the lumbosacral sympathetic chain and the lumbar splanchnic nerves of the cat – Langley revisited. J Auton Nerv Syst (in press)

Baron R, Jänig W, McLachlan EM (1985b) The afferent and sympathetic components of the lumbar spinal outflow to the colon and pelvic organs in the cat: III. The colonic nerves, incorporating an analysis of the components of the whole lumbar prevertebral outflow. J Comp Neurol (in press)

Båth E, Lindblad LE, Wallin BG (1981) Effects of dynamic and static neck suction on muscle nerve sympathetic activity, heart rate and blood pressure in man. J Physiol (Lond) 311:551–564

Beck PW, Handwerker HO, Zimmermann M (1974) Nervous outflow from the cat's foot during noxious radiant heat stimulation. Brain Res 67:373–386

Bell C (1983) Vasodilator neurons supplying skin and skeletal muscle of the limbs. J Auton Nerv Syst 7:257–262

Bell C, Jänig W, Kümmel H, Xu H (1985) Differentiation of vasodilator and sudomotor responses in the cat paw pad to preganglionic sympathetic stimulation. J Physiol (Lond) 364 (in press)

Bernard C (1851) Influence du grand sympathique sur la sensibilité et sur la calorification. C R Soc Biol (Paris) 3:163–164

Bessou P, Perl ER (1969) Response of cutaneous sensory units with unmyelinated fibers to nocious stimuli. J Neurophysiol 32:1025–1043

Bini G, Hagbarth KE, Hynninen P, Wallin BG (1980a) Thermoregulatory and rhythm-generating mechanisms governing the sudomotor and vasoconstrictor outflow in human cutaneous nerves. J Physiol (Lond) 306:537–552

Bini G, Hagbarth KE, Hynninen P, Wallin BG (1980b) Regional similarities and differences in thermoregulatory vaso- and sudomotor tone. J Physiol (Lond) 306:553–565
Bini G, Hagbarth KE, Wallin BG (1981) Cardiac rhythmicity of skin sympathetic activity recorded from peripheral nerves in man. J Auton Nerv Syst 4:17–24
Bishop GH, Heinbecker P (1932) A functional analysis of the cervical sympathetic nerve supply to the eye. Am J Physiol 100:519–532
Blackman JG (1974) Function of autonomic ganglia. In: Hubbard JI (ed) The peripheral nervous system. Plenum, New York, pp 257–276
Blair DA, Glover WE, Greenfield ADM, Roddie IC (1959) Excitation of cholinergic vasodilator nerves in human skeletal muscles during emotional stress. J Physiol (Lond) 148:633–647
Blake DW, Korner PI (1982) Effects of ketamine and althesin anesthesia on baroreceptor-heart rate reflex and hemodynamics of intact and pontine rabbits. J Auton Nerv Syst 5:145–154
Blake DW, Blomberg PA, Korner PI (1982) Effect of ketamine, althesin and thiopentone on the valsalva-constrictor and heart rate reflexes of the rabbit. J Auton Nerv Syst 5:291–301
Blumberg H, Jänig W (1982) Changes in unmyelinated fibers including sympathetic postganglionic fibers of a skin nerve after peripheral neuroma formation. J Auton Nerv Syst 6:173–183
Blumberg H, Jänig W (1983a) Enhancement of resting activity in postganglionic vasoconstrictor neurones following shortlasting repetitive activation of preganglionic axons. Pflügers Arch 396:89–94
Blumberg H, Jänig W (1983b) Changes of reflexes in vasoconstrictor neurons supplying the cat hindlimb following chronic nerve lesions: a model for studying mechanisms of reflex sympathetic dystrophy? J Auton Nerv Syst 7:399–411
Blumberg H, Jänig W (1985) Reflex patterns in postganglionic vasoconstrictor neurons following nerve lesions. J Auton Nerv Syst (in press)
Blumberg H, Jänig W, Rieckmann C, Szulczyk P (1980) Baroreceptor and chemoreceptor reflexes in postganglionic neurones supplying skeletal muscle and hairy skin. J Auton Nerv Syst 2:223–240
Blumberg H, Hilbers K, Jänig W (1983) Viscero-sympathetic reflexes in postganglionic neurones supplying skin and skeletal muscle in brain-intact cats. Naunyn-Schmiedeberg's Arch Pharmacol 322 Suppl R 69
Bopp P (1954) Schwanzfunktion bei Wirbeltieren. Rev Suisse Zool 61:83–151
Bonica JJ (1953) The management of pain. Lea and Febiger, Philadelphia; reprinted by Honji Shoji Company, Tokyo 1980
Bonica JJ (1979) Causalgia and other reflex sympathetic dystrophies. In: Bonica JJ, Liebeskind JC, Albe-Fessard DG (eds) Advances in pain research and therapy, vol 3. Raven, New York, pp 141–166
Brender D, Webb-Peploe MM (1969) Influence of carotid baroreceptors on different components of the vascular system. J Physiol (Lond) 205:257–274
Brooks CM (1933) Reflex activation of the sympathetic system in the spinal cat. Am J Physiol 106:251–266
Brooks CM (1935) The reaction of chronic spinal animals to hermorrhage. Am J Physiol 114:30–39
Brown GE, Adson AW (1929) Physiologic effects of thoracic and of lumbar sympathetic ganglionectomy or section of the trunk. Arch Neurol Psychiat 22:322–357
Burn JH (1938) Sympathetic vasodilator fibres. Physiol Rev 18:137–153
Calof AL, Jones RB, Roberts WJ (1981) Sympathetic modulation of mechanoreceptor sensitivity in frog skin. J Physiol (Lond) 310:481–499
Calvelo MG, Abboud FM, Ballard DR, Abdel-Sayed W (1970) Reflex vascular responses to stimulation of chemoreceptors with nicotine and cyanide. Activation of adrenergic constriction in muscle and noncholinergic dilatation in dog's paw. Circ Res 27: 259–276

Cannon WB (1929) Bodily changes in pain, hunger, fear and rage, 2nd edn. Appleton, New York
Cannon WB (1939) The wisdom of the body. Norton, New York
Chernetski KE (1964) Sympathetic enhancement of peripheral sensory input in the frog. J Neurophysiol 27:493–515
Coggeshall RE, Galbraith SL (1978) Categories of axons in mammalian rami communicantes, part II. J Comp Neurol 181:349–360
Coggeshall RE, Hancock MB, Applebaum ML (1976) Categories of axons in mammalian rami communicantes. J Comp Neurol 167:105–124
Coote JH (1978) Somatic sources of afferent input as factors in aberrant autonomic sensory and motor function. In: Korr IM (ed) The neurobiological mechanisms in manipulative therapy. Plenum, New York, pp 91–127
Coote JH, MacLeod VH (1974a) The influence of bulbospinal monoaminergic pathways on sympathetic nerve activity. J Physiol (Lond) 241:453–475
Coote JH, MacLeod VH (1974b) Evidence for the involvement in the baroreceptor reflex of a descending inhibitory pathway. J Physiol (Lond) 241:477–496
Daly MdeB, Scott MJ (1962) An analysis of the primary cardiovascular reflex effects of stimulation of the carotid body chemoreceptors in the dog. J Physiol (Lond) 162:555–573
Dampney RAL (1981) Functional organization of central cardiovascular pathways. Clin Exp Pharmacol Physiol 8:241–259
Delius W, Hagbarth KE, Hongell A, Wallin BG (1972a) General characteristics of sympathetic activity in human muscle nerves. Acta Physiol Scand 84:65–81
Delius W, Hagbarth KE, Hongell A, Wallin BG (1972b) Manoeuvres affecting sympathetic outflow in human muscle nerves. Acta Physiol Scand 84:82–94
Delius W, Hagbarth KE, Hongell A, Wallin BG (1972c) Manoeuvres affecting sympathetic outflow in human skin nerves. Acta Physiol Scand 84:177–186
Dembowsky K, Lackner K, Czachurski J, Seller H (1981) Tonic catecholaminergic inhibition of the spinal somato-sympathetic reflexes originating in the ventrolateral medulla oblongata. J Auton Nerv Syst 3:277–290
Devor M (1983) Nerve pathophysiology and mechanisms in pain in causalgia. J Auton Nerv Syst 7:371–384
Dusser de Barenne JG (1931) L'influence du système nerveux autonome sur la sensibilité de la peau. J Psychol Norm Pathol 28:177–182
Ectors L (1941) Contribution à l'étude des réactions pilomotrices. Arch Int Physiol 51:443–455
Eldred E, Schnitzlein HN, Buchwald J (1960) Response of muscle spindles to stimulation of the sympathetic trunk. Exp Neurol 2:13–25
Eliasson S, Folkow B, Lindgren P, Uvnäs B (1951) Activation of sympathetic vasodilator nerves to the skeletal muscles in the cat by hypothalamic stimulation. Acta Physiol Scand 23:333–351
Ellison GD, Zanchetti A (1973) Diffuse and specific activation of sympathetic cholinergic fibers of the cat. Am J Physiol 225:142–149
Fitzgerald M (1979) The spread of sensitization of polymodal nociceptors in the rabbit from nearby injury and by antidromic nerve stimulation. J Physiol (Lond) 297: 207–216
Folkow B (1955) Nervous control of the blood vessels. Physiol Rev 35:629–663
Folkow B, Neil E (1971) Circulation. Oxford University Press, New York
Foreman RD, Wurster RD (1973) Localization and functional characteristics of descending sympathetic spinal pathways. Am J Physiol 225:212–217
Frankel HL, Mathias CJ (1976) The cardiovascular system in tetraplegia and paraplegia. In: Vinken PJ, Bruyn GW (eds) Handbook of clinical neurology, vol 26. North-Holland, Amsterdam, pp 313–333
Freeman B, Rowe M (1981) The effect of sympathetic nerve stimulation on responses of cutaneous Pacinian corpuscles in the cat. Neurosci Lett 22:145–150

Gilman AG, Goodman LS, Gilman A (eds) (1980) The pharmacological basis of therapeutics, 6th edn. Macmillan, New York
Golenhofen K (1971) Physiology of skin and muscle circulation. Symposia angiologica santoriana, 3rd international Symposium, part II. Karger, Basel, pp 97–118
Golenhofen K, Hildebrandt G (1957) Psychische Einflüsse auf die Muskeldurchblutung. Pflügers Arch 263:637–646
Golenhofen K, Blair DA, Seidel W (1961) Zur Natur affektiver Muskeldurchblutungssteigerungen beim Menschen. Pflügers Arch 272:223–236
Graham BH, Lioy F (1973) Histaminergic vasodilatation in the hindlimb of the dog. Pflügers Arch 342:307–318
Granit R (1975) The functional role of the muscle spindles – facts and hypotheses. Brain 98:531–556
Granit R (1981) Comments on history of motor control. In: Brooks VB (ed) Motor control, part I. American Physiological Society, Bethesda, pp 1–16 (Handbook of physiology, section 1, The nervous system, vol II)
Gregor M, Jänig W (1977) Effects of systemic hypoxia and hypercapnia on cutaneous and muscle vasoconstrictor neurones to the cat's hindlimb. Pflügers Arch 368:71–81
Gregor M, Jänig W, Riedel W (1976) Repsonse pattern of cutaneous postganglionic neurones to the hindlimb on spinal cord heating and cooling in the cat. Pflügers Arch 363:135–140
Gregor M, Jänig W, Wiprich L (1977) Cardiac and respiratory rhythmicities in cutaneous and muscle vasoconstrictor neurones to the cat's hindlimb. Pflügers Arch 370: 299–302
Grewe W, Jänig W, Kümmel H, Varma S (1982) Effects of central thermal stimuli on activity in cutaneous and muscle vasoconstrictor neurones supplying the cat hindlimb. Pflügers Arch 394:R 53
Grillner S (1975) Locomotion in vertebrates: central mechanisms and reflex interaction. Physiol Rev 55:247–304
Grillner S (1981) Control of locomotion in bipedes, tetrapodes, and fish. In: Brooks VB (ed) Motor control, part 2. American Physiological Society, Bethesda, pp 1179–1236 (Handbook of physiology, section I, The nervous system, vol II)
Grosse M, Jänig W (1976) Vasoconstrictor and pilomotor fibres in skin nerves to the cat's tail. Pflügers Arch 361:221–229
Guth L, Bernstein JJ (1961) Selectivity in the re-establishment of synapses in the superior cervical sympathetic ganglion of the cat. Exp Neurol 4:59–69
Guttmann L (1976a) Spinal cord injuries. Comprehensive management and research, 2nd edn. Blackwell, Oxford
Guttmann L (1976b) Spinal shock. In: Vinken PJ, Bruyn GW (eds) Injuries of the spine and spinal cord. Part II. Handbook of clinical neurology, vol 26. North-Holland, Amsterdam, pp 243–262
Hagbarth KE, Hallin RG, Hongell A, Torebjörk HE, Wallin BG (1972) General characteristics of sympathetic activity in human skin nerves. Acta Physiol Scand 84:164–176
Hainsworth R, Linden RJ (1979) Reflex control of vascular capacitance. Int Rev Physiol 18:67–124
Hall M (1841) On the diseases and derangements of the nervous system. Balliere, London
Hallin RG, Torebjörk HE (1974) Single unit sympathetic activity in human skin nerves during rest and various manoeuvres. Acta Physiol Scand 92:303–317
Hallin RG, Wiesenfeld-Hallin Z (1983) Does sympathetic activity modify afferent inflow at the receptor level in man? J Auton Nerv Syst 7:391–397
Head H, Riddoch G (1917) The autonomic bladder, excessive sweating and some other reflex conditions, in gross injuries of the spinal cord. Brain 40:188–263
Hellmann K (1963) The effect of temperature changes on the isolated pilomotor muscles. J Physiol (Lond) 169:621–629

Hensel H (1981) Thermoreception and temperature regulation. Monographs of the Physiological Society No. 38. Academic, London
Hilton SM, Spyer KM (1980) Central nervous regulation of vascular resistance. Annu Rev Physiol 42:399–411
Hoffmeister B, Hussels W, Jänig W (1978) Long-lasting discharge of postganglionic neurones to skin and muscle of the cat's hindlimb after repetitive activation of preganglionic axons in the lumbar sympathetic trunk. Pflügers Arch 376:15–20
Hökfelt T, Johansson O, Ljungdahl A, Lundberg JM, Schultzberg M (1980) Peptidergic neurones. Nature 284:515–521
Homma S (ed) (1976) Understanding the stretch reflex. Prog Brain Res 44:1–507
Horeyseck G, Jänig W (1973) Somato-sympathetic reflexes in cutaneous and muscular sympathetic fibres in decerebrate cats. Pflügers Arch 343: R 65
Horeyseck G, Jänig W (1974a) Reflexes in postganglionic fibres within skin and muscle nerves after mechanical non-noxious stimulation of skin. Exp Brain Res 20:115–123
Horeyseck G, Jänig W (1974b) Reflexes in postganglionic fibres within skin and muscle nerves after noxious stimulation of skin. Exp Brain Res 20:125–134
Horeyseck G, Jänig W (1974c) Reflex activity in postganglionic fibres within skin and muscle nerves elicited by somatic stimuli in chronic spinal cats. Exp Brain Res 21: 155–168
Horeyseck G, Jänig W, Kirchner F, Thämer V (1972) Activation of muscle vasodilator neurons by hypothalamic stimulation. Brain Res 48:394–396
Horeyseck G, Jänig W, Kirchner F, Thämer V (1976) Activation and inhibition of muscle and cutaneous postganglionic neurones to hindlimb during hypothalamically induced vasoconstriction and atropine-sensitive vasodilation. Pflügers Arch 361: 231–240
Horwitz TS, Kaufman AI (1979) Effects of baroreceptor activation on spontaneous activity in the sweat glands and nictitating membrane of the cat. Brain Res 178: 411–424
Hunsperger RW (1965) Neurophysiologische Grundlagen des affektiven Verhaltens. Bull Schweiz Akad Med Wiss 21:8–22
Hunt CC (1960) The effect of sympathetic stimulation on mammalian muscle spindles. J Physiol (Lond) 151:332–341
Hunt CC, Jami L, Laporte Y (1982) Effects of stimulating the lumbar sympathetic trunk on cat hindlimb muscle spindles. Arch Ital Biol 120:371–384
Illert M, Gabriel M (1972) Descending pathways in the cervical cord of cats affecting blood pressure and sympathetic activity. Pflügers Arch 335:109–124
Illert M, Seller H (1969) A descending sympathoinhibitory tract in the ventrolateral column of the cat. Pflügers Arch 313:343–360
Iriki M, Kozawa E (1976) Patterns of differentiation in various sympathetic efferents induced by hypoxic and by central thermal stimulation in decerebrated rabbits. Pflügers Arch 362:101–108
Iriki M, Walther OE, Pleschka K, Simon E (1971a) Regional cutaneous and visceral sympathetic activity during asphyxia in the anesthetized rabbit. Pflügers Arch 322: 167–182
Iriki M, Pleschka K, Walther OE, Simon E (1971b) Hypoxia and hypercapnia in asphyctic differentiation of regional sympathetic activity in the anesthetized rabbit. Pflügers Arch 328:91–102
Jänig W (1975) Central organization of somato-sympathetic reflexes in vasoconstrictor neurones. Brain Res 87:305–312
Jänig W (1979) Reciprocal reaction patterns of sympathetic subsystems with respect to various afferent inputs. In: Brooks CMcC, Koizumi K, Sato A (eds) Integrative functions of the autonomic nervous system. University of Tokyo Press, Tokyo. Elsevier, Amsterdam, pp 263–274
Jänig W (1982) An idea about the organization of the lumbar sympathetic outflow supplying skeletal muscle and skin. Proc Aust Physiol Pharmacol Soc 13:47–58

Jänig W (1984) Vasoconstrictor systems supplying skeletal muscle, skin, and viscera. Clin Exp Hypertens (A) 6:329–346

Jänig W, Kümmel H (1977) Functional discrimination of postganglionic neurones to the cat's hindpaw with respect to the skin potentials recorded from the hairless skin. Pflügers Arch 371:217–225

Jänig W, Kümmel H (1981) Organization of the sympathetic innervation supplying the hairless skin of the cat's paw. J Auton Nerv Syst 3:215–230

Jänig W, McLachlan E (1984) On the fate of sympathetic and sensory neurons projecting into a neuroma of the superficial peroneal nerve in the cat. J Comp Neurol 225:302–311

Jänig W, Räth B (1977) Electrodermal reflexes in the cat's paws elicited by natural stimulation of skin. Pflügers Arch 369:27–32

Jänig W, Räth B (1980) Effects of anaesthetics on reflexes elicited in the sudomotor system by stimulation of Pacinian corpuscles and of cutaneous nociceptors. J Auton Nerv Syst 2:1–14

Jänig W, Schmidt RF (1970) Single unit responses in the cervical sympathetic trunk upon somatic nerve stimulation. Pflügers Arch 314:199–216

Jänig W, Spilok N (1978) Functional organization of the sympathetic innervation supplying the hairless skin of the hindpaws in chronic spinal cats. Pflügers Arch 377:25–31

Jänig W, Szulczyk P (1979) Conduction velocity in spinal descending pathways of baro- and chemoreceptor reflex. J Auton Nerv Syst 1:149–160

Jänig W, Szulczyk P (1980) Functional properties of lumbar preganglionic neurones. Brain Res 186:115–131

Jänig W, Szulczyk P (1981) The organization of lumbar preganglionic neurons. J Auton Nerv Syst 3:177–191

Jänig W, Schmidt RF, Zimmermann M (1968) Single unit responses and the total afferent outflow from the cat's foot pad upon mechanical stimulation. Exp Brain Res 6:100–115

Jänig W, Sato A, Schmidt RF (1972) Reflexes in postganglionic cutaneous fibres by stimulation of group I to group IV somatic afferents. Pflügers Arch 331:244–256

Jänig W, Kümmel H, Wiprich L (1980) Respiratory rhythmicities in vasoconstrictor and sudomotor neurones supplying the cat's hindlimb. In: Koepchen HD, Hilton SM, Trzebski A (eds) Central interaction between respiratory and cardiovascular control systems. Springer, Berlin Heidelberg New York, pp 128–135

Jänig W, Krauspe R, Wiedersatz G (1982) Transmission of impulses from pre- to postganglionic vasoconstrictor and sudomotor neurons. J Auton Nerv Syst 6:95–106

Jänig W, Krauspe R, Wiedersatz G (1983) Reflex activation of postganglionic vasoconstrictor neurones supplying skeletal muscle by stimulation of arterial chemoreceptors via non-nicotinic synaptic mechanisms in sympathetic ganglia. Pflügers Arch 396:95–100

Jänig W, Sundlöf G, Wallin BG (1983) Discharge patterns of sympathetic neurons supplying skeletal muscle and skin in man and cat. J Auton Nerv Syst 7:239–256

Jankowska E, Lundberg A (1981) Interneurones in the spinal cord. Trends Neurosci 4: 230–233

Kandel ER (1976) Cellular basis of behavior. Freeman, San Francisco

Karim F, Araneda G, Hainsworth R (1980) The influence of perfusate temperature on the responses of a superficial vein in the carotid baroreceptor reflex in dogs. Pflügers Arch 383:79–85

Kaufman A, Koizumi K (1971) Spontaneous and reflex activity of single units in lumbar white rami. In: Kao FF, Koizumi K, Vasalle M (eds) Research in physiology. Aulo Gaggi, Bologna, pp 469–481

Kendrick E, Öberg B, Wennergren G (1972) Vasoconstrictor fibre discharge to skeletal muscle, kidney, intestine and skin at varying levels of arterial baroreceptor activity in the cat. Acta Physiol Scand 85:464–476

Kirchheim HR (1976) Systemic arterial baroreceptor reflexes. Physiol Rev 56:100–176

Koepchen HP, Hilton SM, Trzebski A (eds) (1980) Central interaction between respiratory and cardiovascular control systems. Springer, Berlin Heidelberg New York

Koizumi K, Brooks CM (1972) The integration of autonomic reactions: a discussion of autonomic reflexes, their control and their association with somatic reactions. Ergeb Physiol 67:1–68

Korner PI (1971) Integrative neural cardiovascular control. Physiol Rev 51:312–367

Korner PI (1979) Central nervous control of autonomic cardiovascular function. In: Barne RM (ed) The cardiovascular system, vol I. The heart. Handbook of physiology. American Physiological Society, Bethesda, pp 691–739

Korner PI, Uther JB (1975) Reflex autonomic control of heart rate and peripheral blood flow. Brain Res 87:293–303

Kosterlitz HW, Thompson JW, Wallis DI (1964) The compound action potential in the nerve supplying the medial smooth muscle of the nictitating membrane of the cat. J Physiol (Lond) 171:426–433

Kuhn RA (1950) Functional capacity of the isolated human spinal cord. Brain 73:1–51

Kümmel H (1983) Activity in sympathetic neurons supplying skin and skeletal muscle in spinal cats. J Auton Nerv Syst 7:319–327

Kümmel H, Xu H (1983) Viscero-sympathetic reflexes in postganglionic neurones supplying the hindlimb of chronic spinal cats. Naunyn-Schmiedeberg's Arch Pharmacol 322:R69

Kuntz A (1951) Afferent innervation of peripheral blood vessels through sympathetic trunks. South Med J 44:674–678

Kuntz A, Farnsworth DI (1931) Distribution of afferent fibers via the sympathetic trunks and gray communicating rami to the brachial and lumbosacral plexuses. J Comp Neurol 53:389–399

Kuntz A, Saccomanno G (1942) Afferent conduction from extremities through dorsal root fibers via sympathetic trunks. Arch Surg 45:606–612

Kuo DC, Hisamitsu T, de Groat WC (1984) A sympathetic projection from sacral paravertebral ganglia to the pelvic nerve and to postganglionic nerves on the surface of the urinary bladder and large intestine of the cat. J Comp Neurol 226:76–86

Langley JN (1891a) On the course and connections of the secretory fibres supplying the sweat glands of the feet of the cat. J Physiol (Lond) 12:347–374

Langley JN (1891b) Note on the connection with nerve cells of the vaso motor nerves for the feet. J Physiol (Lond) 12:375–377

Langley JN (1892) On the origin from the spinal cord of the cervical and upper thoracic sympathetic fibres with some observations on white and grey rami communicantes. Philos Trans R Soc (Lond) (Biol) B 183:85–124

Langley JN (1894) The arrangement of the sympathetic nervous system, based chiefly on observations upon pilo-motor nerves. J Physiol (Lond) 15:176–244

Langley JN (1894/5) Further observations on the secretory and vaso-motor fibres of the foot of the cat, with notes on other sympathetic nerve fibres. J Physiol (Lond) 17:296–314

Langley JN (1903) Das sympathische und verwandte nervöse System der Wirbeltiere (autonomes nervöses System). Ergeb Physiol 2/II:818–872

Langley JN, Anderson HK (1894) On reflex action from sympathetic ganglia. J Physiol (Lond) 16:410–440

Langley JN, Sherrington CS (1891) On pilo-motor nerves. J Physiol (Lond) 12:278–291

Leitner JM, Perl ER (1964) Receptors supplied by spinal nerves which respond to cardiovascular changes and adrenaline. J Physiol (Lond) 175:254–274

Lewis T, Marvin HM (1927) Observations upon a pilomotor reaction in response to faradism. J Physiol (Lond) 64:87–106
Leyhausen P (1979) Katzen, eine Verhaltenskunde, 5th edn. Parey, Berlin
Lisander B (1970) Factors influencing the autonomic component of the defence reaction. Acta Physiol Scand 78 (Suppl 351):1–42
Little R, Öberg B (1975) Circulatory responses to stimulation of the carotid body chemoreceptors in the cat. Acta Physiol Scand 93:34–51
Litwin MS (1962) Postsympathectomy neuralgia. Arch Surg 84:591–595
Livingston WK (1943) Pain mechanisms. A physiologic interpretation of causalgia and its related states. Macmillan, New York; Plenum, New York, 1976
Loewenstein WR (1956) Modulation of cutaneous mechanoreceptors by sympathetic stimulation. J Physiol (Lond) 132:40–60
Loewenstein WR, Altamirano-Orrego R (1956) Enhancement of activity in a Pacinian corpuscle by sympathomimetic agents. Nature 178:1292–1293
Loewy AD (1982) Descending pathways to the sympathetic preganglionic neurons. Prog Brain Res 57:267–277
Loewy AD, Neil JJ (1981) The role of descending monaminergic systems in the central control of blood pressure. Fed Proc 40:2778–2785
Lovén C (1866) Über die Erweiterung von Arterien in Folge einer Nervenerregung. Ber Verh königl sächs Ges Wiss Math phys Classe 18:85–110
Lundberg A (1975) Control of spinal mechanisms from the brain. In: Brady RC (ed) The basic neurosciences. Raven, New York, pp 253–265 (The nervous system, vol I)
Lundberg A (1979a) Integration in a propriospinal motor centre controlling the forelimb in the cat. In: Asanuma H, Wilson VJ (eds) Integration in the nervous system. Igaku Shoin, Tokyo, pp 47:64
Lundberg A (1979b) Multisensory control of spinal reflex pathways. In: Granit R, Pompeiano O (eds) Reflex control of posture and movement. Prog Brain Res 50: 11–28
Lundberg JM (1981) Evidence for coexistence of vasoactive intestinal polypeptide (VIP) and acetylcholine in neurons of cat exocrine glands. Morphological, biochemical and functional studies. Acta Physiol Scand, Suppl 496:1–57
Lynn B (1977) Cutaneous hyperalgesia. Br Med Bull 33:103–108
Mancia G (1975) Influence of carotid baroreceptors on vascular responses to carotid chemoreceptors in the dog. Circ Res 36:270–276
Mancia G, Baccelli G, Zanchetti A (1972) Hemodynamic responses to different emotional stimuli in the cat: patterns and mechanisms. Am J Physiol 223:925–933
Mannard A, Polosa C (1973) Analysis of background firing of single sympathetic preganglionic neurons of cat cervical nerve. J Neurophysiol 36:398–408
McLachlan EM, Hirst GDS (1980) Some properties of preganglionic neurons in upper thoracic spinal cord of the cat. J Neurophysiol 43:1251–1265
McLachlan EM, Jänig W (1983) The cell bodies of origin of sympathetic and sensory axons in some skin and muscle nerves of the cat hindlimb. J Comp Neurol 214: 115–130
Mellander S, Johansson B (1968) Control of resistance, exchange and capacitance functions in the peripheral circulation. Pharmacol Rev 20:117–196
Muller KJ, Nicholls JG, Stent GS (eds) (1981) Neurobiology of the leech. Cold Spring Harbor Laboratory, New York
Murray JG, Thompson JW (1957) The occurrence and function of collateral sprouting in the sympathetic nervous system of the cat. J Physiol (Lond) 135:133–162
Nilsson BY (1972) Effects of sympathetic stimulation on mechanoreceptors of cat vibrissae. Acta Physiol Scand 85:390–397
Ninomiya I, Irisawa H (1975) Non-uniformity of the sympathetic nerve activity in response to baroreceptor inputs. Brain Res 87:313–322

Ninomiya I, Irisawa A, Nisimaru N (1973) Nonuniformity of sympathetic nerve activity to the skin and kidney. Am J Physiol 224:256–264
Njå A, Purves D (1977) Specific innervation of guinea-pig superior cervical ganglion cells by preganglionic fibres arising from different levels of the spinal cord. J Physiol (Lond) 264:565–583
Nyström TG (1949) Lumbar sympathectomy: late results in chronic obliterative arterial diseases of the legs. Arch Chir Scand, Suppl 142:1–127
Pelletier CL (1972) Circulatory responses to graded stimulation of the carotid chemoreceptors in the dog. Circ Res 31:431–443
Pelletier CL, Shepherd JT (1972) Venous responses to stimulation of carotid chemoreceptors by hypoxia and hypercapnia. Am J Physiol 223:97–103
Pelletier CL, Shepherd JT (1975) Effect of hypoxia on vascular responses to the carotid baroreflex. Am J Physiol 228:331–336
Perl ER, Kumazawa T, Lynn B, Kenins P (1976) Sensitization of high threshold receptors with unmyelinated (C) afferent fibers. In: Iggo A, Ilyinsky OB (eds) Somatosensory and visceral receptor mechanisms. Prog Brain Res 43:263–276
Peter W, Riedel W (1982) Neurogenic non-adrenergic cutaneous vasodilatation elicited by hypothalamic thermal stimulation in dogs. Pflügers Arch 395:115–120
Pette H (1927) Das Problem der wechselseitigen Beziehungen zwischen Sympathicus und Sensibilität. Dtsch Z Nervenheilk 100:143–164
Pick J (1970) The autonomic nervous system. Lippincott, Philadelphia
Pierce JP, Roberts WJ (1981) Sympathetically induced changes in the responses of guard hair and type II receptors in the cat. J Physiol (Lond) 314:411–428
Pleschka K (1984) Control to tongue blood flow in regulation of heat loss in mammals. Rev Physiol Biochem Pharmacol 100:75–120
Polosa C (1968) Spontaneous activity of sympathetic preganglionic neurons. Can J Physiol Pharmacol 46:887–896
Polosa C, Wyszogrodski I (1973) The inhibition of sympathetic preganglionic neurons by somatic afferents. Can J Physiol Pharmacol 51:29–38
Purves D, Lichtman JW (1978) Formation and maintenance of synaptic connections in autonomic ganglia. Physiol Rev 58:821–862
Purves D, Wigston DJ (1983) Neural units in the superior cervical ganglion of the guinea-pig. J Physiol (Lond) 334:169–178
Richards RL (1967) Causalgia. Arch Neurol 16:339–350
Richter DW (1982) Generation and maintenance of the respiratory rhythm. J Exp Biol 100:93–107
Richter DW, Ballantyne D (1983) A three phase theory about the basic respiratory pattern generator. In: Schläfke ME, Koepchen HP, See WR (eds) Central neurone environment and the control systems of breathing and circulation. Springer, Berlin Heidelberg New York Tokyo, pp 164–174
Riddoch G, Head H (1917) The reflex functions of the completely divided spinal cord in man, compared with those associated with less severe lesions. Brain 40:264–402
Roberts WJ, Elardo SM (1985) Sympathetic activation of A-delta nociceptors. Somatosensory Research (in press)
Rosell S (1980) Neuronal control of microvessels. Ann Rev Physiol 42:359–371
Rosell S, Belfrage E (1979) Blood circulation in adipose tissue. Physiol Rev 59:1078–1104
Rowell LB (1981) Active neurogenic vasodilatation in man. In: Vanhoutte PM, Leusen I (eds) Vasodilatation. Raven, New York, pp 1–17
Sahs AL, Fulton JF (1940) Somatic and autonomic reflexes in spinal monkeys. J Neurophysiol 3:258–268
Santini M (1976) Towards a theory of sympathetic-sensory coupling: the primary sensensory neuron as a feedback target of the sympathetic terminal. In: Zottermann Y (ed) Sensory functions of the skin in primates. Pergamon, Oxford, pp 15–35
Santini M, Ibata Y, Pappas GD (1971) The fine structure of the sympathetic axons within the Pacinian corpuscle. Brain Res 33:279–287

Sato A (1972) The relative involvement of different reflex pathways in somatosympathetic reflexes, analyzed in spontaneously active single preganglionic sympathetic units. Pflügers Arch 333:70–81
Sato A, Schmidt RF (1973) Somatosympathetic reflexes: afferent fibers, central pathways, discharge characteristics. Physiol Rev 53:916–947
Schiff JD (1974) Role of the sympathetic innervation of the Pacinian corpuscle. J Gen Physiol 63:601–608
Schlaefke ME (1981) Central chemosensitivity: a respiratory drive. Rev Physiol Biochem Pharmacol 90:171–244
Schönung W, Wagner H, Simon E (1972) Neurogenic vasodilatatory component in the thermoregulatory skin blood flow response of the dog. Naunyn-Schmiedeberg's Arch Pharmacol 273:230–241
Seller H (1973) The discharge pattern of single units in thoracic and lumbar white rami in relation to cardiovascular events. Pflügers Arch 343:317–330
Sherrington CS (1906) The integrative action of the nervous system. Yale University Press, New Haven
Simon E (1974) Temperature regulation: the spinal cord as a site of extrahypothalamic thermoregulatory functions. Rev Physiol Biochem Pharmacol 71:1–76
Simon E, Iriki M (1971) Sensory transmission of spinal heat and cold sensitivity in ascending spinal neurons. Pflügers Arch 328:103–120
Skok VI (1973) Physiology of autonomic ganglia. Igaku Shoin, Tokyo
Skok VI, Ivanov AY (1983) What is the ongoing activity of sympathetic neurons? J Auton Nerv Syst 7:263–270
Sonnenschein RR, Weissmann ML (1978) Sympathetic vasomotor outflow to hindlimb muscles of the cat. Am J Physiol 235:H482–487
Spray DC (1974) Characteristics, specificity, and efferent control of frog cutaneous cold receptors. J Physiol (Lond) 237:15–38
Spyer KM (1981) Neural organization and control of the baroreceptor reflex. Rev Physiol Biochem Pharmacol 88:23–124
Strickland JH, Calhoun ML (1983) The integumentary system of the cat. Amer J Vet Res 24:1018–1029
Sunderland S (1978) Nerves and nerve injuries, 2nd edn. Livingstone, Edinburgh
Sundlöf G, Wallin BG (1977) The variability of muscle nerve sympathetic activity in resting recumbent man. J Physiol (Lond) 272:383–397
Sundlöf G, Wallin BG (1978) Effect of lower body negative pressure on human muscle nerve sympathetic activity. J Physiol (Lond) 278:525–532
Takeuchi T, Manning JW (1971) Muscle cholinergic dilators in the sinus baroreceptor response in cats. Circ Res 29:350–357
Takeuchi T, Manning JW (1973) Hypothalamic mediation of sinus baroreceptor-evoked muscle cholinergic dilator response. Am J Physiol 224:1280–1287
Tedeschi RE, Sherman S, de Sanctis N, Davidheiser S, Schainbaum J (1971) Effect of carotid sinus baroreceptor stimulation on blood pressure and sympathetic outflow. Am J Physiol 221:405–412
Thauer R (1970) Thermosensitivity of the spinal cord. In: Hardy JD, Gagge APh, Stolwijk JAJ (eds) Physiological and behavioral temperature regulation. Thomas, Springfield, pp 472–492
Tracy GD, Cockett FB (1957) Pain in the lower limb after sympathectomy. Lancet I: 12–14
Uvnäs B (1954) Sympathetic vasodilator outflow. Physiol Rev 34:608–618
Uvnäs B (1960) Sympathetic vasodilator system and blood flow. Physiol Rev, Suppl 4:69–76
Uvnäs B (1967) Cholinergic vasodilator innervation to skeletal muscles. Circ Res 20/21 (Suppl 1):83–90
Vallbo ÅB, Hagbarth KE, Torebjörk HE, Wallin BG (1979) Somatosensory, proprioceptive, and sympathetic activity in human peripheral nerves. Physiol Rev 59:919–957

Vanhoutte PM (1980) Physical factors of regulation. In: Bohr DF, Somlyo AP, Sparks HV (eds) Vascular smooth muscle. American Physiological Society, Bethesda, pp 443–474 (Handbook of physiology, section 2, The cardiovascular system, vol 2)
von Brücke ET (1932) Einflüsse des vegetativen Nervensystems auf Vorgänge innerhalb des animalischen Systems. Ergeb Physiol 34:220–252
Wallin BG (1981) New aspects of sympathetic function in man. Butterworths, London, pp 145–167
Wallin BG, Eckberg DL (1982) Sympathetic transients caused by abrupt alterations of carotid baroreceptor activity in humans. Am J Physiol 242:H185–H190
Wallin BG, König U (1976) Changes of skin nerve sympathetic activity during induction of anaesthesia with thiopentone in man. Brain Res 103:157–160
Wallin BG, Sundlöf G, Delius W (1975) The effect of carotid sinus nerve stimulation on muscle and skin nerve sympathetic activity in man. Pflügers Arch 358:101–110
Wang GH (1964) The neural control of sweating. The University of Wisconsin Press, Madison
Wang GH, Chun RWS (1967) Sweating under different ambient temperatures in normal, striatal and thalamic cats. Arch Ital Biol 105:379–392
Weissman ML, Rubinstein EH, Sonnenschein RR (1976) Vascular responses to short-term systemic hypoxia, hypercapnia, and asphyxia in the cat. Am J Physiol 230: 595–601
Wennergren G (1975) Aspects of central integrative and efferent mechanisms in cardiovascular reflex control. Acta Physiol Scand, Suppl 428:1–53
Wolthuis RA, Bergman SA, Nicogossian AE (1974) Physiological effects of locally applied reduced pressure in man. Physiol Rev 54:566–595
Zanchetti A, Baccelli G, Mancia G, Ellison GD (1972) Emotion and the cardiovascular system in the cat. In: Physiology, emotion and psychosomatic illness, Ciba Foundation Symposium 8 (new series) Elsevier, Amsterdam, pp 201–219

Author Index

Subject Index

GPSR Compliance

The European Union's (EU) General Product Safety Regulation (GPSR) is a set of rules that requires consumer products to be safe and our obligations to ensure this.

If you have any concerns about our products, you can contact us on ProductSafety@springernature.com

In case Publisher is established outside the EU, the EU authorized representative is:

Springer Nature Customer Service Center GmbH
Europaplatz 3
69115 Heidelberg, Germany

Batch number: 10620881

Printed by Printforce, the Netherlands